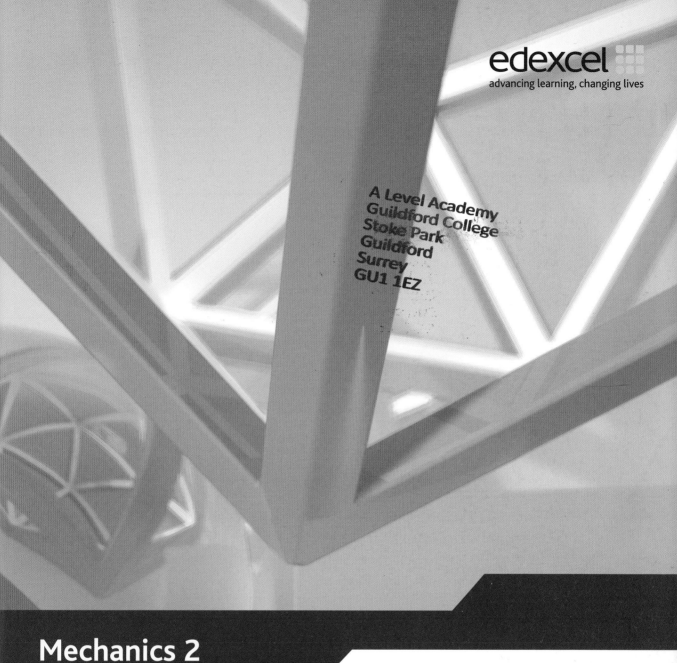

edexcel
advancing learning, changing lives

A Level Academy
Guildford College
Stoke Park
Guildford
Surrey
GU1 1EZ

Mechanics 2

Edexcel AS and A-level Modular Mathematics

Susan Hooker
Michael Jennings
Jean Littlewood
Bronwen Moran
Laurence Pateman

Contents

About this book   iv

1   Kinematics of a particle moving in a straight line or plane   1
    1.1 The constant acceleration formulae for a projectile in a vertical plane   2
    1.2 Calculus for a particle moving in a straight line with acceleration that varies with time   11
    1.3 Calculus, with vectors, for a particle moving in a plane   19

2   Centres of mass   31
    2.1 Finding the centre of mass of a set of particles in a straight line   32
    2.2 Finding the centre of mass of a set of particles in a plane   34
    2.3 Finding the centre of mass of a standard uniform plane lamina   39
    2.4 Finding the centre of mass of a composite lamina   44
    2.5 Finding the centre of mass of a framework   50
    2.6 Solving problems involving a lamina in equilibrium   53

3   Work, energy and power   63
    3.1 Calculating the work done by a force when its point of application moves   64
    3.2 Calculating the kinetic energy and potential energy of a moving particle   68
    3.3 The principle of conservation of mechanical energy and the work–energy principle   72
    3.4 Calculating the power developed by an engine   77

    Review Exercise 1   86

4   Collisions   100
    4.1 The impulse–momentum principle   101
    4.2 Conservation of linear momentum and Newton's Law of Restitution   104
    4.3 Newton's Law of Restitution and collisions of a particle with a smooth plane   111
    4.4 Problems involving successive impacts   114
    4.5 Finding a change in energy due to an impact or application of an impulse   121

5   Statics of rigid bodies   131
    5.1 Moments of forces acting on a body   132
    5.2 Rigid bodies in equilibrium   134
    5.3 Rigid bodies under the action of three non-parallel forces   139
    5.4 Rigid bodies in limiting equilibrium   141

Review Exercise 2   151

Practice exam paper   161

Answers   164

Index   170

About this book

This book is designed to provide you with the best preparation possible for your Edexcel M2 unit examination:

- This is Edexcel's own course for the GCE specification.
- Written by senior examiners.
- The LiveText CD-ROM in the back of the book contains even more resources to support you through the unit.
- A matching M2 revision guide is also available.

Finding your way around the book

Brief chapter overview and 'links' to underline the importance of mathematics: to the real world, to your study of further units and to your career

Every few chapters, a review exercise helps you consolidate your learning

Detailed contents list shows which parts of the M2 specification are covered in each section

After completing this chapter you should be able to:

1 solve problems involving the motion of projectiles
2 solve problems involving motion in a straight line when acceleration varies with time
3 use calculus and vectors to solve problems involving motion in two dimensions.

Kinematics of a particle moving in a straight line or plane

A particle moving in a vertical plane is sometimes called a **projectile**. You could use projectile motion to model the flight of a golf ball. In this chapter you will make the following **modelling assumptions**:

- Air resistance can be ignored.
- Objects can be modelled as particles.
- The force due to gravity is uniform and acts vertically downwards.
- Objects only move in a vertical plane. There is no side-to-side movement.

Usually the only force acting on a projectile is gravity, which is modelled as being constant. When other forces act on a projectile they can vary with time. As this rocket travels into space, fuel is burnt and the mass of the rocket decreases with time. The acceleration of the rocket is not constant but increases with time.

Contents

About this book

1 Kinematics of a particle moving in a straight line or plane
   1.1 The constant acceleration formulae for a projectile in a vertical p
   1.2 Calculus for a particle moving in a straight line with acceleration
   1.3 Calculus, with vectors, for a particle moving in a plane

2 Centres of mass
   2.1 Finding the centre of mass of a set of particles in a straight line
   2.2 Finding the centre of mass of a set of particles in a plane
   2.3 Finding the centre of mass of a standard uniform plane lamina
   2.4 Finding the centre of mass of a composite lamina
   2.5 Finding the centre of mass of a framework
   2.6 Solving problems involving a lamina in equilibrium

3 Work, energy and power
   3.1 Calculating the work done by a force when its point of applicati
   3.2 Calculating the kinetic energy and potential energy of a moving
   3.3 The principle of conservation of mechanical energy and the wo
   3.4 Calculating the power developed by an engine

Review Exercise 1

4 Collisions
   4.1 The impulse-momentum principle
   4.2 Conservation of linear momentum and Newton's Law of Restitu
   4.3 Newton's Law of Restitution and collisions of a particle with a s
   4.4 Problems involving successive impacts
   4.5 Finding a change in energy due to an impact or application of an

5 Statics of rigid bodies
   5.1 Moments of forces acting on a body
   5.2 Rigid bodies in equilibrium
   5.3 Rigid bodies under the action of three non-parallel forces
   5.4 Rigid bodies in limiting equilibrium

Review Exercise

Whenever a numerical value of g is required, take $g = 9.8\,\mathrm{m\,s^{-1}}$.

1 A stone was thrown with velocity $20\,\mathrm{m\,s^{-1}}$ at an angle of elevation of 30° from the top of a vertical cliff. The stone moved freely under gravity and reached the sea 5 s after it was thrown. Find

   a the vertical height above the sea from which the stone was thrown,

   b the horizontal distance covered by the stone from the instant when it was thrown until it reached the sea,

   c the magnitude and direction of the velocity of the stone when it reached the sea.

2 A darts player throws darts at a dart board which hangs vertically. The motion of a dart is modelled as that of a particle moving freely under gravity. The darts move in a vertical plane which is perpendicular to the plane of the dart board. A dart is thrown horizontally with speed $12.6\,\mathrm{m\,s^{-1}}$. It hits the board at a point which is 10 cm below the level from which it was thrown.

   a Find the horizontal distance from the point where the dart was thrown to the dart board.

The darts player moves his position. He now throws a dart from a point which is at a horizontal distance of 2.5 m from the dart board. He throws the dart at an angle of elevation α to the horizontal where $\tan \alpha = \frac{1}{14}$. The dart hits the board at a point which is at the same level as the point from which it was thrown.

   b Find the speed with which the dart was thrown.

3 A particle is projected with velocity $(8\mathbf{i} + 10\mathbf{j})\,\mathrm{m\,s^{-1}}$, where $\mathbf{i}$ and $\mathbf{j}$ are unit vectors horizontally and vertically respectively, from a point O at the top of a cliff and moves freely under gravity. Six seconds after projection, the particle strikes the sea at the point 3. Calculate

   a the horizontal distance between O and 3,

   b the vertical distance between O and 3.

Each section begins with a statement of what is covered in the section

5.1 You can calculate the moment of a force acting on a body.

The moment of a force measures the turning effect of the force on the body on which it is acting.

■ The moment of a force F about a point P is the product of the magnitude of the force and the perpendicular distance of the line of action of the force from the point P.

The moment of the force is measured in newton-metres (Nm). When describing the turning effect of the force you need to consider its magnitude and the sense of the rotation (clockwise or anticlockwise).

Example 1

The diagram shows three forces acting on a lamina. Find the sum of the moments of these forces about P.

Method 1

Because the line of action of the 7 N force is not perpendicular to the distance you have been given, you need to start by finding the perpendicular distance.

$\circ 7 \times 1.2\sin 65° = 7.612\ldots$ Nm
$\circ 5 \times 0.8 = 4$ Nm
Total of moments
$= 7.612\ldots - 4$
$= 3.61$ Nm clockwise.

Method 2

Rather than find the perpendicular distance, you might find it easier to resolve the 7 N force into two components parallel and perpendicular to the given distance.

$\circ 7\sin 65° \times 1.2 = 7.612\ldots$ Nm
$\circ 5 \times 0.8 = 4$ Nm
Total of moments
$= 3.61$ Nm clockwise.

You get the same answer with both methods.

Exercise 5A

Find the sum of the moments about P of the forces shown in the following questions.

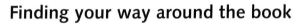

Concise learning points

Step-by-step worked examples – they are model solutions and include examiners' hints

Past examination questions are marked 'E'

Each section ends with an exercise – the questions are carefully graded so they increase in difficulty and gradually bring you up to standard

Each chapter has a different colour scheme, to help you find the right chapter quickly

Each chapter ends with a mixed exercise and a summary of key points.

At the end of the book there is an examination-style paper.

LiveText software

The LiveText software gives you additional resources: Solutionbank and Exam café. Simply turn the pages of the electronic book to the page you need, and explore!

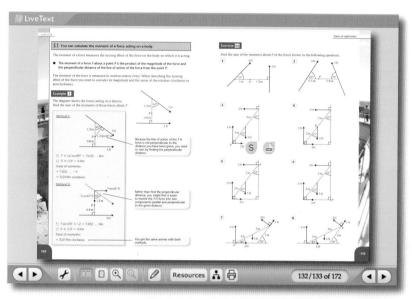

Unique Exam café feature:

- Relax and prepare – revision planner; hints and tips; common mistakes
- Refresh your memory – revision checklist; language of the examination; glossary
- Get the result! – fully worked examination-style paper with chief examiner's commentary

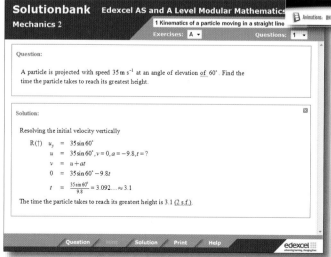

Solutionbank

- Hints and solutions to every question in the textbook
- Solutions and commentary for all review exercises and the practice examination paper

Published Pearson Education Limited, a company incorporated in England and Wales, having its registered office at Edinburgh Gate, Harlow, Essex, CM20 2JE. Registered company number: 872828

Text © Susan Hooker, Michael Jennings, Jean Littlewood, Bronwen Moran, Laurence Pateman 2009

12
10 9 8

British Library Cataloguing in Publication Data is available from the British Library on request.

ISBN 978 0 435519 179

Edited by Harry Smith and Susan Gardner
Typeset by Tech-Set Ltd, Gateshead
Illustrated by Tech-Set Ltd, Gateshead
Cover design by Christopher Howson
Picture research by Chrissie Martin
Index by Indexing Specialists (UK) Ltd
Cover photo/illustration © Science Photo Library/Laguna Design
Printed in China (CTPS/08)

Acknowledgements

The author and publisher would like to thank the following individuals and organisations for permission to reproduce photographs:

NASA/Marshall Space Flight Center p**1**; Photoshot/Lang Congliu p**34**; Shutterstock/Ernita p**63**, Alamy/Mitchell Gunn p**100**; Photoshot/Art Foxhall p**131**

Every effort has been made to contact copyright holders of material reproduced in this book. Any omissions will be rectified in subsequent printings if notice is given to the publishers.

Disclaimer

This Edexcel publication offers high-quality support for the delivery of Edexcel qualifications.
Edexcel endorsement does not mean that this material is essential to achieve any Edexcel qualification, nor does it mean that this is the only suitable material available to support any Edexcel qualification. No endorsed material will be used verbatim in setting any Edexcel examination/assessment and any resource lists produced by Edexcel shall include this and other appropriate texts.
Copies of official specifications for all Edexcel qualifications may be found on the Edexcel website – www.edexcel.com

After completing this chapter you should be able to:

1 solve problems involving the motion of projectiles
2 solve problems involving motion in a straight line when acceleration varies with time
3 use calculus and vectors to solve problems involving motion in two dimensions.

Kinematics of a particle moving in a straight line or plane

1

A particle moving in a vertical plane is sometimes called a **projectile**. You could use projectile motion to model the flight of a golf ball. In this chapter you will make the following **modelling assumptions**:

- Air resistance can be ignored.
- Objects can be modelled as particles.
- The force due to gravity is uniform and acts vertically downwards.
- Objects only move in a vertical plane. There is no side-to-side movement.

Usually the only force acting on a projectile is gravity, which is modelled as being constant. When other forces act on a projectile they can vary with time. As this rocket travels into space, fuel is burnt and the mass of the rocket decreases with time. The acceleration of the rocket is not constant but increases with time.

1.1 You can use the constant acceleration formulae for a projectile moving in a vertical plane.

When a particle is projected with speed u, at an angle α to the horizontal, it will move along a symmetric curve.

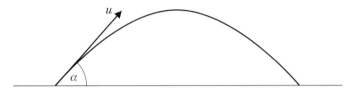

The initial speed u is called the **speed of projection** of the particle.
The angle α is called the **angle of projection** or **angle of elevation** of the particle.

The initial velocity of the projectile can be resolved into two components.

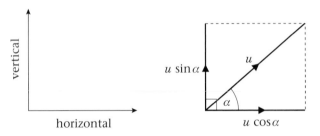

The horizontal component of the initial velocity is $u \cos \alpha$.
The **vertical component** of the initial velocity is $u \sin \alpha$.

■ **The vertical motion of the projectile is motion with constant acceleration $g = 9.8\,\text{m s}^{-2}$.**
You can use the constant acceleration formulae from book M1.

■ **The horizontal motion of a projectile is motion with constant speed. You can use the formula distance = speed × time.**

The distance from the point from which the particle was projected to the point where it strikes the horizontal plane is called the **range**.

The time the particle takes to move from its point of projection to the point where it strikes the horizontal plane is called the **time of flight** of the projectile.

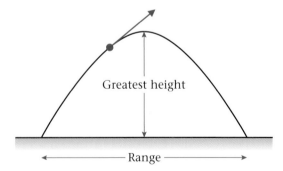

Example **1**

A particle P is projected from a point O on a horizontal plane with speed $28\,\text{m s}^{-1}$ and with angle of elevation $30°$. After projection, the particle moves freely under gravity until it strikes the plane at a point A. Find

a the greatest height above the plane reached by P,

b the time of flight of P,

c the distance OA.

Resolving the velocity of projection horizontally and vertically

$R(\rightarrow)$ $\quad u_x = 28 \cos 30° = 24.248\ldots$

$R(\uparrow)$ $\quad u_y = 28 \sin 30° = 14$

Resolve the velocity of projection horizontally and vertically:

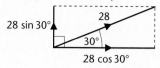

a Taking the upwards direction as positive

$R(\uparrow)$ $\quad u = 14, v = 0, a = -9.8, s = ?$

$v^2 = u^2 + 2as$

$0^2 = 14^2 - 2 \times 9.8 \times s$

$s = \dfrac{14^2}{2 \times 9.8} = 10$

The greatest height above the plane reached by P is 10 m.

At the highest point the vertical component of the velocity is zero.

The vertical motion is motion with constant acceleration. You can use the formulae you learnt in Chapter 2 of book M1.

b The particle strikes the plane when the vertical displacement is zero.

$R(\uparrow)$ $\quad s = 0, u = 14, a = -9.8, t = ?$

$s = ut + \frac{1}{2}at^2$

$0 = 14t - 4.9t^2$

$\quad = t(14 - 4.9t)$

$t = 0$

or $\quad t = \dfrac{14}{4.9} = 2.857\ldots \approx 2.9$

The time of flight is 2.9 s, to 2 significant figures.

When the particle strikes the plane, it is at the same height (zero) as when it started.

$t = 0$ corresponds to the point from which P was projected and can be ignored.

The value for g was given to 2 significant figures. You should not give your answer to a greater accuracy than the data used, so you should round your answer to 2 significant figures.

c $R(\rightarrow)$ $\quad$ distance $=$ speed $\times$ time

$\quad\quad\quad\quad = 28 \cos 30° \times 2.857\ldots$

$\quad\quad\quad\quad = 69.282\ldots \approx 69$

$OA = 69$ m, to 2 significant figures

There is no horizontal acceleration.

To find the total distance moved horizontally you use the time of flight found in part **b**. You should use the unrounded value (2.857...), not the rounded value (2.9). Otherwise you will make rounding errors.

Example 2

A ball is thrown horizontally, with speed $20\,\text{m s}^{-1}$, from the top of a building which is $30\,\text{m}$ high. Find

a the time the ball takes to reach the ground,

b the distance between the bottom of the building and the point where the ball strikes the ground.

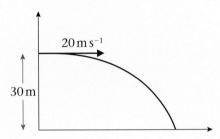

Resolving the velocity of projection horizontally and vertically

$$R(\rightarrow) \quad u_x = 20$$
$$R(\downarrow) \quad u_y = 0$$

Model the ball as a particle and the ground as a horizontal plane and ignore air resistance.

As the ball is thrown horizontally the initial horizontal component of the velocity is $20\,\text{m s}^{-1}$ and the initial vertical component is zero.

a Taking the downwards direction as positive

$$R(\downarrow) \quad u = 0, \ s = 30, \ a = 9.8, \ t = ?$$
$$s = ut + \tfrac{1}{2}at^2$$
$$30 = 0 + 4.9 \times t^2$$
$$t^2 = \frac{30}{4.9} \Rightarrow t = \sqrt{\left(\frac{30}{4.9}\right)} = 2.474\ldots \approx 2.5$$

The ball takes $2.5\,\text{s}$, to 2 significant figures, to reach the ground.

The vertical motion is with constant acceleration. The final velocity v is not involved here, so you choose the formula without v.

The downwards direction is positive, so the value for a is positive.

b $R(\rightarrow)$ distance $=$ speed $\times$ time
$$= 20 \times 2.474\ldots = 49.487\ldots$$
$$\approx 49$$

The ball hits the ground $49\,\text{m}$, to 2 significant figures, from the bottom of the building.

The horizontal motion is with constant velocity.

Example 3

A particle is projected from a point O with speed $V\,\mathrm{m\,s^{-1}}$ and at an angle of elevation of θ, where $\tan\theta = \frac{4}{3}$. The point O is 42.5 m above a horizontal plane. The particle strikes the plane, at a point A, 5 s after it is projected.

a Show that $V = 20$.

b Find the distance between O and A.

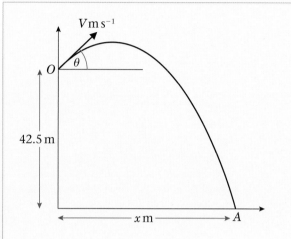

Start by drawing a diagram.

Resolving the velocity of projection horizontally and vertically

$$R(\rightarrow) \quad u_x = V\cos\theta = \tfrac{3}{5}V$$

$$R(\uparrow) \quad u_y = V\sin\theta = \tfrac{4}{5}V$$

You will need $\sin\theta$ and $\cos\theta$ to resolve the initial velocity. When you know $\tan\theta$ you can draw a triangle to find $\cos\theta$ and $\sin\theta$.

$\tan\theta = \frac{4}{3}$

$\sin\theta = \frac{4}{5}$

$\cos\theta = \frac{3}{5}$

a Taking the upwards direction as positive

$$R(\uparrow) \quad s = -42.5,\ u = \tfrac{4}{5}V,\ g = -9.8,\ t = 5$$

$$s = ut + \tfrac{1}{2}at^2$$

You use the formula $s = ut + \frac{1}{2}at^2$ to obtain an equation in V, which you solve.

$$-42.5 = \tfrac{4}{5}V \times 5 - 4.9 \times 25$$

$$4V = 4.9 \times 25 - 42.5 = 80$$

$$V = \tfrac{80}{4} = 20,\ \text{as required.}$$

b Let the horizontal distance moved be x m

$$R(\rightarrow)\ \text{distance} = \text{speed} \times \text{time}$$

$$x = \tfrac{3}{5}V \times 5 = 3V = 60$$

You use the value of V found in part **a** to find the horizontal distance moved by the particle.

Using Pythagoras' Theorem

$$OA^2 = 42.5^2 + 60^2 = 5406.25$$

$$OA = \sqrt{5406.25} = 73.527\ldots \approx 74$$

The distance between O and A is 74 m, to 2 significant figures.

Example 4

A particle is projected from a point O with speed $35\ \mathrm{m\,s^{-1}}$ at an angle of elevation of $30°$. The particle moves freely under gravity.

Find the length of time for which the particle is $15\ \mathrm{m}$ or more above O.

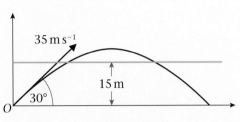

Resolving the initial velocity vertically

$R(\uparrow)\quad u_y = 35\sin 30° = 17.5$

$s = 15,\ u = 17.5,\ a = -9.8,\ t = ?$

$s = ut + \tfrac{1}{2}at^2$

$15 = 17.5t - 4.9t^2$

$4.9t^2 - 17.5t + 15 = 0$

Multiplying by 10

$49t^2 - 175t + 150 = 0$

$(7t - 10)(7t - 15) = 0$

$t = \tfrac{10}{7},\ \tfrac{15}{7}$

$\tfrac{15}{7} - \tfrac{10}{7} = \tfrac{5}{7}$

The particle is $15\ \mathrm{m}$ or more above O for $\tfrac{5}{7}$ s.

The particle is 15 m above O twice. First on the way up and then on the way down.

In this example the horizontal component of the initial velocity is not used.

You form a quadratic equation in t to find the two times when the particle is 15 m above O. Between these two times, the particle will be more than 15 m above O.

These factors are not easy to spot so you could use the formula for solving a quadratic equation.

You could also give this answer as a decimal to 2 significant figures, 0.71 s.

Example 5

A particle is projected from a point with speed u at an angle of elevation α and moves freely under gravity. When the particle has moved a horizontal distance x, its height above the point of projection is y.

a Show that $y = x\tan\alpha - \dfrac{gx^2}{2u^2}(1 + \tan^2\alpha)$

A particle is projected from a point A on a horizontal plane, with speed $28\ \mathrm{m\,s^{-1}}$ at an angle of elevation α. The particle passes through a point B, which is at a horizontal distance of $32\ \mathrm{m}$ from A and at a height of $8\ \mathrm{m}$ above the plane.

b Find the two possible values of α, giving your answers to the nearest degree.

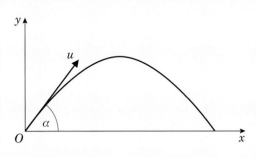

a $\quad R(\rightarrow) \qquad u_x = u \cos \alpha$

$\quad R(\uparrow) \qquad u_y = u \sin \alpha$

For the horizontal motion

$\qquad$ distance = speed $\times$ time

$\qquad x = u \cos \alpha \times t \qquad$ ➊

For the vertical motion, taking upwards as positive

$\qquad R(\uparrow) \quad s = ut + \frac{1}{2}at^2$

$\qquad y = u \sin \alpha \times t - \frac{1}{2}gt^2 \qquad$ ➋

Rearranging ➊ to make t the subject of the formula

$\qquad t = \dfrac{x}{u \cos \alpha} \qquad$ ➌

Substituting ➌ into ➋

$\qquad y = u \sin \alpha \times \dfrac{x}{u \cos \alpha} - \dfrac{1}{2}g\left(\dfrac{x}{u \cos \alpha}\right)^2$

Using $\tan \alpha = \dfrac{\sin \alpha}{\cos \alpha}$ and $\dfrac{1}{\cos \alpha} = \sec \alpha$,

$\qquad y = x \tan \alpha - \dfrac{gx^2}{2u^2} \sec^2 \alpha$

Using $\sec^2 \alpha = 1 + \tan^2 \alpha$,

$\qquad y = x \tan \alpha - \dfrac{gx^2}{2u^2}(1 + \tan^2 \alpha)$, as required.

b $\quad$ Using the result in **a** with $u = 28$, $x = 32$, $y = 8$

and $g = 9.8$

$\qquad 8 = 32 \tan \alpha - 6.4(1 + \tan^2 \alpha)$

Rearranging as a quadratic in $\tan \alpha$

$\qquad 6.4 \tan^2 \alpha - 32 \tan \alpha + 14.4 = 0$

Divide all terms by 1.6

$\qquad 4 \tan^2 \alpha - 20 \tan \alpha + 9 = 0$

$\qquad (2 \tan \alpha - 1)(2 \tan \alpha - 9) = 0$

$\qquad \tan \alpha = \frac{1}{2}, \frac{9}{2}$

$\qquad \alpha = 27°$ and $77°$, to the nearest degree

Resolve the velocity of projection horizontally and vertically.

If the upwards direction is taken as positive, the vertical acceleration is $-g$.

You have obtained two equations, labelled ➊ and ➋. Both equations contain t and the result you have been asked to show has no t in it. You must eliminate t using substitution.

The identity $\sec^2 x = 1 + \tan^2 x$ can be found in Chapter 6 of book C3. The M2 specification assumes that you have studied modules M1, C1, C2 and C3.

You substitute the values given in part **b** into the result given in part **a**.

You could use the quadratic formula to solve this equation.

There are two possible angles of elevation for which the particle will pass through B. This sketch illustrates the two paths.

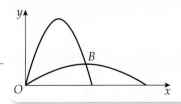

Example 6

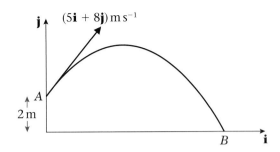

A ball is struck by a racket at a point A which is 2 m above horizontal ground. Immediately after being struck, the ball has velocity $(5\mathbf{i} + 8\mathbf{j})\,\mathrm{m\,s}^{-1}$, where $\mathbf{i}$ and $\mathbf{j}$ are unit vectors horizontally and vertically respectively. After being struck, the ball travels freely under gravity until it strikes the ground at the point B, as shown in the diagram above. Find

a the greatest height above the ground reached by the ball,

b the speed of the ball as it reaches B,

c the angle the velocity of the ball makes with the ground as the ball reaches B.

a Taking the upwards direction as positive

$R(\uparrow)$ $u = 8, v = 0, a = -9.8, s = ?$

$v^2 = u^2 + 2as$

$0^2 = 8^2 - 2 \times 9.8 \times s$

$s = \dfrac{64}{19.6} = 3.265... \approx 3.3$

The greatest height above the ground reached by the ball is $(2 + 3.3)\,\mathrm{m} = 5.3\,\mathrm{m}$, to 2 significant figures.

b The horizontal component of the velocity of the ball at B is $5\,\mathrm{m\,s}^{-1}$.

The vertical component of the velocity of the ball at B is given by

$R(\uparrow)$ $s = -2, u = 8, a = -9.8, v = ?$

$v^2 = u^2 + 2as$

$= 8^2 + 2 \times (-9.8) \times (-2) = 103.2$

The speed at B is given by

$v^2 = 5^2 + 103.2 = 128.2$

$v = \sqrt{128.2} \approx 11$

The speed of the ball as it reaches B is $11\,\mathrm{m\,s}^{-1}$, to 2 significant figures.

c The angle is given by

$\tan\theta = \dfrac{\sqrt{103.2}}{5} \approx 2.032 \Rightarrow \theta \approx 64°$

The angle the velocity of the ball makes with the ground as the ball reaches B is $64°$, to the nearest degree.

The velocity of projection has been given as a vector in terms of $\mathbf{i}$ and $\mathbf{j}$. The horizontal component is 5 and the vertical component is 8.

3.3 m is the greatest height above the point of projection. You need to add 2 m to find the height above the ground.

The horizontal motion is motion with constant speed, so the horizontal component of the velocity never changes.

There is no need to find the square root of 103.2 at this point, as you need v^2 in the next stage of the calculation.

As the ball reaches B, its velocity has two components as shown below.

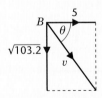

The magnitude (speed) and direction of the velocity are found using trigonometry and Pythagoras' Theorem.

Exercise 1A

Whenever a numerical value of g is required, take $g = 9.8\,\mathrm{m\,s^{-2}}$.

1 A particle is projected with speed $35\,\mathrm{m\,s^{-1}}$ at an angle of elevation of $60°$. Find the time the particle takes to reach its greatest height.

2 A ball is projected from a point $5\,\mathrm{m}$ above horizontal ground with speed $18\,\mathrm{m\,s^{-1}}$ at an angle of elevation of $40°$. Find the height of the ball above the ground $2\,\mathrm{s}$ after projection.

3 A stone is projected horizontally from a point above horizontal ground with speed $32\,\mathrm{m\,s^{-1}}$. The stone takes $2.5\,\mathrm{s}$ to reach the ground. Find

 a the height of the point of projection above the ground,

 b the distance from the point on the ground vertically below the point of projection to the point where the stone reached the ground.

4 A projectile is launched from a point on horizontal ground with speed $150\,\mathrm{m\,s^{-1}}$ at an angle of $10°$ to the horizontal. Find

 a the time the projectile takes to reach its highest point above the ground,

 b the range of the projectile.

5 A particle is projected from a point O on a horizontal plane with speed $20\,\mathrm{m\,s^{-1}}$ at an angle of elevation of $45°$. The particle moves freely under gravity until it strikes the ground at a point X. Find

 a the greatest height above the plane reached by the particle,

 b the distance OX.

6 A ball is projected from a point A on level ground with speed $24\,\mathrm{m\,s^{-1}}$. The ball is projected at an angle θ to the horizontal where $\sin\theta = \frac{4}{5}$. The ball moves freely under gravity until it strikes the ground at a point B. Find

 a the time of flight of the ball,

 b the distance from A to B.

7 A particle is projected with speed $21\,\mathrm{m\,s^{-1}}$ at an angle of elevation α. Given that the greatest height reached above the point of projection is $15\,\mathrm{m}$, find the value of α, giving your answer to the nearest degree.

8 A particle is projected horizontally from a point A which is $16\,\mathrm{m}$ above horizontal ground. The projectile strikes the ground at a point B which is at a horizontal distance of $140\,\mathrm{m}$ from A. Find the speed of projection of the particle.

9 A particle P is projected from the origin with velocity $(12\mathbf{i} + 24\mathbf{j})\,\mathrm{m\,s^{-1}}$, where $\mathbf{i}$ and $\mathbf{j}$ are horizontal and vertical unit vectors respectively. The particle moves freely under gravity. Find

 a the position vector of P after $3\,\mathrm{s}$,

 b the speed of P after $3\,\mathrm{s}$.

10 A stone is thrown with speed $30 \, \text{m s}^{-1}$ from a window which is $20 \, \text{m}$ above horizontal ground. The stone hits the ground $3.5 \, \text{s}$ later. Find

   a the angle of projection of the stone,

   b the horizontal distance from the window to the point where the stone hits the ground.

11 A ball is thrown from a point O on horizontal ground with speed $u \, \text{m s}^{-1}$ at an angle of elevation of θ, where $\tan \theta = \frac{3}{4}$. The ball strikes a vertical wall which is $20 \, \text{m}$ from O at a point which is $3 \, \text{m}$ above the ground. Find

   a the value of u,

   b the time from the instant the ball is thrown to the instant that it strikes the wall.

12

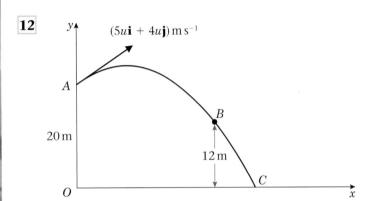

[*In this question, the unit vectors* **i** *and* **j** *are in a vertical plane,* **i** *being horizontal and* **j** *being vertical.*]

A particle P is projected from a point A with position vector $20\mathbf{j} \, \text{m}$ with respect to a fixed origin O. The velocity of projection is $(5u\mathbf{i} + 4u\mathbf{j}) \, \text{m s}^{-1}$. The particle moves freely under gravity, passing through a point B, which has position vector $(k\mathbf{i} + 12\mathbf{j}) \, \text{m}$, where k is a constant, before reaching the point C on the x-axis, as shown in the figure above. The particle takes $4 \, \text{s}$ to move from A to B. Find

   a the value of u,

   b the value of k,

   c the angle the velocity of P makes with the x-axis as it reaches C.

13

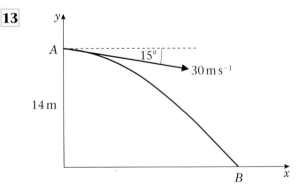

A stone is thrown from a point A with speed $30 \, \text{m s}^{-1}$ at an angle of $15°$ below the horizontal. The point A is $14 \, \text{m}$ above horizontal ground. The stone strikes the ground at the point B, as shown in the figure above. Find

   a the time the stone takes to travel from A to B,   **b** the distance AB.

14 A particle is projected from a point with speed $21\,\mathrm{m\,s^{-1}}$ at an angle of elevation α and moves freely under gravity. When the particle has moved a horizontal distance $x\,\mathrm{m}$, its height above the point of projection is $y\,\mathrm{m}$.

  a Show that $y = x\tan\alpha - \dfrac{x^2}{90\cos^2\alpha}$.

  b Given that $y = 8.1$ when $x = 36$, find the value of $\tan\alpha$.

15 A projectile is launched from a point on a horizontal plane with initial speed $u\,\mathrm{m\,s^{-1}}$ at an angle of elevation α. The particle moves freely under gravity until it strikes the plane. The range of the projectile is $R\,\mathrm{m}$.

  a Show that the time of flight of the particle is $\dfrac{2u\sin\alpha}{g}$ seconds.

  b Show that $R = \dfrac{u^2\sin 2\alpha}{g}$.

  c Deduce that, for a fixed u, the greatest possible range is when $\alpha = 45°$.

  d Given that $R = \dfrac{2u^2}{5g}$, find the two possible values of the angle of elevation at which the projectile could have been launched.

16 A particle is projected from a point on level ground with speed $u\,\mathrm{m\,s^{-1}}$ and angle of elevation α. The maximum height reached by the particle is $42\,\mathrm{m}$ above the ground and the particle hits the ground $196\,\mathrm{m}$ from its point of projection.

  Find the value of α and the value of u.

1.2 You can use calculus for a particle moving in a straight line with acceleration that varies with time.

In M1 you used formulae like $s = ut + \frac{1}{2}at^2$ for particles moving in straight lines. These formulae can only be used when a particle is moving with constant acceleration. If the acceleration of the particle varies, you must use calculus.

This particle is moving in a straight line with acceleration a, displacement x and velocity v. The relationship between these three variables is shown in the diagram below.

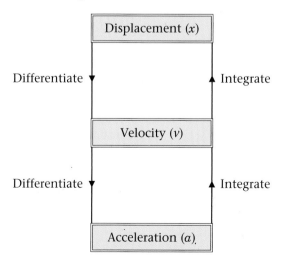

■ Velocity is the rate of change of displacement with time. To find the velocity from the displacement, you differentiate with respect to time.

■ Acceleration is the rate of change of velocity with time. To find the acceleration from the velocity you differentiate with respect to time.

■ To obtain the velocity from the acceleration, you integrate with respect to time.

■ To obtain the displacement from the velocity, you integrate with respect to time.

Using symbols:

$$v = \frac{dx}{dt}$$

$$a = \frac{dv}{dt} = \frac{d^2x}{dt^2}$$

$$v = \int a\,dt$$

$$x = \int v\,dt$$

When you integrate, it is important that you remember to include a constant of integration. The constant of integration often represents the initial displacement or initial velocity of a particle. Many questions include information which enables you to find the value of this constant.

Example **7**

A particle P is moving on the x-axis. At time t seconds, the displacement x metres from O is given by $x = t^4 - 32t + 12$. Find

a the speed of P when $t = 3$,

b the value of t for which P is instantaneously at rest,

c the magnitude of the acceleration of P when $t = 1.5$.

a $x = t^4 - 32t + 12$

$v = \dfrac{dx}{dt} = 4t^3 - 32$

When $t = 3$,

$v = 4 \times 3^3 - 32 = 76$

The speed of P when $t = 3$ is $76\,\text{m s}^{-1}$.

You find the velocity by differentiating the displacement.

To find the velocity when $t = 3$, you substitute $t = 3$ into the expression.

b $v = 4t^3 - 32 = 0$

$t^3 = \dfrac{32}{4} = 8$

$t = 2$

The particle is at rest when $v = 0$. You substitute $v = 0$ into your expression for v and solve the resulting equation to find t.

c $v = 4t^3 - 32$

$a = \dfrac{dv}{dt} = 12t^2$

When $t = 1.5$,

$a = 12 \times 1.5^2 = 27$

The magnitude of the acceleration of P when

$t = 1.5$ is $27\,\text{m s}^{-2}$.

You find the acceleration by differentiating the velocity.

Example 8

A particle is moving on the x-axis. At time $t = 0$, the particle is at the point where $x = 5$. The velocity of the particle at time t seconds (where $t \geqslant 0$) is $(6t - t^2)\,\text{m s}^{-1}$. Find

a the acceleration of the particle when $t = 2$,

b the acceleration of the particle when $t = 4$,

c an expression for the displacement of the particle from O at time t seconds,

d the distance of the particle from its starting point when $t = 6$.

a $v = 6t - t^2$

$a = \dfrac{dv}{dt} = 6 - 2t$

When $t = 2$,

$a = 6 - 2 \times 2 = 2$

When $t = 2$, the acceleration of the particle is $2\,\text{m s}^{-1}$ directed away from the origin O.

You find the acceleration by differentiating the velocity.

b When $t = 4$,

$a = 6 - 2 \times 4 = -2$

When $t = 2$, the acceleration of the particle is $2\,\text{m s}^{-1}$ directed towards the origin O.

Acceleration is a vector quantity. When specifying an acceleration, you should give the direction of the acceleration as well as its magnitude.

c $x = \displaystyle\int v\,dt$

$= 3t^2 - \dfrac{t^3}{3} + c$, where c is a constant of integration.

You integrate the velocity to find the displacement. You must remember to add the constant of integration.

When $t = 0$, $x = 5$

$5 = 3 \times 0^2 - \dfrac{0^3}{3} + c = c \Rightarrow c = 5$

This information enables you to find the value of the constant of integration.

The displacement of the particle from O after t seconds is $\left(3t^2 - \dfrac{t^3}{3} + 5\right)\text{m}$.

d Using the result in **c**, when $t = 6$

$x = 3 \times 6^2 - \dfrac{6^3}{3} + 5 = 41$

The displacement from the starting point is

$(41 - 5)\,\text{m} = 36\,\text{m}$.

This calculation shows you that, when $t = 6$ the particle is 41 m from O. When the particle started, it was 5 m from O. So the distance from its starting point is $(41 - 5)\,\text{m}$.

Example 9

A particle P is moving along a straight line. At time $t = 0$, the particle is at a point A and is moving with velocity $8\,\text{m s}^{-1}$ towards a point B on the line, where $AB = 30\,\text{m}$. At time t seconds (where $t \geqslant 0$), the acceleration of P is $(2 - 2t)\,\text{m s}^{-2}$ in the direction $\overrightarrow{AB}$.

a Find an expression, in terms of t, for the displacement of P from A at time t seconds.

b Show that P does not reach B.

c Find the value of t when P returns to A, giving your answer to 3 significant figures.

d Find the total distance travelled by P in the interval between the two instants when it passes through A.

a $\quad v = \int a\,dt = \int (2 - 2t)\,dt$

$\qquad = 2t - t^2 + c$, where c is a constant of integration

When $t = 0$, $v = 8$

$\qquad 8 = 0 + 0 + c \Rightarrow c = 8$

Let s m be the displacement from A at time t seconds

$\qquad v = 8 + 2t - t^2$

$\qquad s = \int v\,dt = \int (8 + 2t - t^2)\,dt$

$\qquad = 8t + t^2 - \dfrac{t^3}{3} + k$, where k is a constant of integration

When $t = 0$, $s = 0$

$\qquad 0 = 0 + 0 - 0 + k \Rightarrow k = 0$

The displacement of P from A at time t seconds is

$\qquad \left(8t + t^2 - \dfrac{t^3}{3}\right)$ m.

b The greatest positive displacement of P occurs when

$\qquad \dfrac{ds}{dt} = v = 0$

From **a** $\quad v = 8 + 2t - t^2 = 0$

$\qquad\qquad t^2 - 2t - 8 = (t - 4)(t + 2) = 0$

$\qquad\qquad t = 4$

When $t = 4$, $s = 8 \times 4 + 4^2 - \dfrac{4^3}{3} = 26\frac{2}{3} < 30$

Hence, P does not reach B.

To obtain the displacement from the acceleration, you have to integrate twice. At each stage you must use the information in the question to find the constant of integration.

The second constant of integration is 0. Even when it seems obvious that a constant has this value, you should show sufficient working to justify the value 0.

At its greatest displacement, the particle changes direction and, instantaneously, its velocity is zero. Alternatively, using calculus, the displacement has its maximum when its derivative is zero. As the derivative of the displacement is the velocity, these two ways of looking at the question are equivalent.

The question states that $t \geqslant 0$, so the second solution $t = -2$ is not needed.

c   At A, $8t + t^2 - \dfrac{t^3}{3} = 0$

$t\left(8 + t - \dfrac{t^2}{3}\right) = 0$

$8 + t - \dfrac{t^2}{3} = 0$

$t^2 - 3t - 24 = 0$

$t = \dfrac{3 + \sqrt{(9 + 4 \times 24)}}{2} \approx 6.62$

> When P returns to A its displacement from its starting point is 0.

> Again you can ignore the negative solution.

d   The total distance moved by A is

$2 \times 26\frac{2}{3} = 53\frac{1}{3}$ m.

> From the working to part **b**, P moves from A to a point $26\frac{2}{3}$ m from A and then back to A.

Example 10

A small metal ball moving in a magnetic field is modelled as a particle P of mass 0.2 kg, moving in a straight line under the action of a single variable force **F** newtons. At time t seconds, the displacement, x metres, of B from A is given by $x = 3 \sin 2t$.

Find the magnitude of **F** when $t = \dfrac{\pi}{6}$.

$x = 3 \sin 2t$

$v = \dfrac{dx}{dt} = 6 \cos 2t$

$a = \dfrac{dv}{dt} = -12 \sin 2t$

When $t = \dfrac{\pi}{6}$,

$a = -12 \sin\left(2 \times \dfrac{\pi}{6}\right) = -12 \sin \dfrac{\pi}{3} = -6\sqrt{3}$

Using **F** = m**a**,

$F = 0.2 \times -6\sqrt{3} = -1.2\sqrt{3}$

The magnitude of **F** when $t = \dfrac{\pi}{6}$ is $1.2\sqrt{3}$

> You find forces using Newton's Second Law, **F** = m**a**. You can find the acceleration by differentiating the displacement twice.

> The M2 specification requires knowledge of the C3 specification. This includes knowledge of differentiating trigonometric, exponential and logarithmic functions. It does not require knowledge of the integration of these functions.

> Approximate answers, such as 2.1 to 2 significant figures, would also be acceptable.

Example 11

A particle P moves on the x-axis. At time t seconds, the velocity of P is $v\,\mathrm{m\,s^{-1}}$ in the direction of x increasing, where v is given by

$$v = \begin{cases} 5t, & 0 \leqslant t < 1, \\ t + \dfrac{4}{t^2}, & 1 \leqslant t \leqslant 3, \\ 3\tfrac{4}{9}, & t < 3. \end{cases}$$

When $t = 0$, P is at the origin O.

a Find the least speed of P in the interval $1 \leqslant t \leqslant 3$.

b Sketch a velocity–time graph to illustrate the motion of P in the interval $0 \leqslant t \leqslant 6$.

c Find the distance of P from O when $t = 6$.

a $v = t + 4t^{-2}$

The least value of v is when

$\dfrac{dv}{dt} = 1 + (-2)4t^{-3} = 0$

$1 - \dfrac{8}{t^3} = 0$

$t^3 = 8 \Rightarrow t = 2$

The least speed of P is $\left(2 + \dfrac{4}{2^2}\right)\mathrm{m\,s^{-1}} = 3\,\mathrm{m\,s^{-1}}$.

To differentiate $\dfrac{4}{t^2}$, you write it as $4t^{-2}$ and use the result $\dfrac{d}{dt}(t^n) = nt^{n-1}$.

You find the least value by using calculus to find the minimum. You differentiate v and equate the result to 0. You can then substitute the resulting value of t to find the least value of v.

b
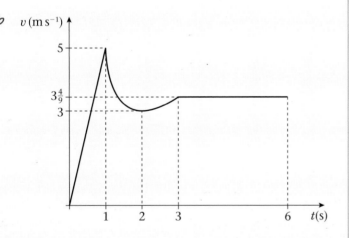

c The distance travelled in the first second is

$$\frac{1}{2} \times 1 \times 5 = 2\frac{1}{2}\,\text{m}$$

For $1 \leqslant t \leqslant 3$

$$s = \int v\,dt$$

$$= \int (t + 4t^{-2})\,dt = \frac{t^2}{2} + \frac{4t^{-1}}{-1} + c$$

$$= \frac{t^2}{2} - \frac{4}{t} + c$$

where c is a constant of integration.

When $t = 1$, $s = 2\frac{1}{2}$

$$2\frac{1}{2} = \frac{1}{2} - \frac{4}{1} + c$$

$$c = 2\frac{1}{2} - \frac{1}{2} + 4 = 6$$

So $\quad s = \frac{t^2}{2} - \frac{4}{t} + 6$

When $t = 3$,

$$s = \frac{3^2}{2} - \frac{4}{3} + 6 = 9\frac{1}{6}$$

From $t = 3$ to $t = 6$ P moves a further

$$\left(3\frac{4}{9} \times 3\right)\text{m} = 10\frac{1}{3}\,\text{m}$$

When $t = 6$, the distance of P from O is

$$9\frac{1}{6} + 10\frac{1}{3}\,\text{m} = 19\frac{1}{2}\,\text{m}.$$

> You can find the distance moved in the first second of motion and in the interval $3 < t \leqslant 6$ using the areas under the graph in these intervals.

> After one second P has moved $2\frac{1}{2}$ m. You use this as the initial value in the interval $1 \leqslant t \leqslant 3$ to find the constant of integration.

> You can also find the distance moved in the interval $1 \leqslant t \leqslant 3$ by calculating the definite integral
> $$\int_1^3 (t + 4t^{-2})\,dt$$
> This represents the area under the velocity–time graph between $t = 1$ and $t = 3$.

Exercise 1B

1 A particle is moving in a straight line. At time t seconds, its displacement, x m, from a fixed point O on the line is given by $x = 2t^3 - 8t$. Find

  a the speed of the particle when $t = 3$,

  b the magnitude of the acceleration of the particle when $t = 2$.

2 A particle P is moving on the x-axis. At time t seconds, the velocity of P is $(8 + 2t - 3t^2)\,\text{m s}^{-1}$ in the direction of x increasing. At time $t = 0$, P is at the point where $x = 4$. Find

  a the magnitude of the acceleration of P when $t = 3$,

  b the distance of P from O when $t = 1$.

3 A particle P is moving on the x-axis. At time t seconds, the acceleration of P is $(16 - 2t)\,\mathrm{m\,s^{-2}}$ in the direction of x increasing. The velocity of P at time t seconds is $v\,\mathrm{m\,s^{-1}}$. When $t = 0$, $v = 6$ and when $t = 3$, $x = 75$. Find

a v in terms of t,

b the value of x when $t = 0$.

4 A particle P is moving on the x-axis. At time t seconds (where $t \geqslant 0$), the velocity of P is $v\,\mathrm{m\,s^{-1}}$ in the direction of x increasing, where $v = 12 - t - t^2$.

Find the acceleration of P when P is instantaneously at rest.

5 A particle is moving in a straight line. At time t seconds, its displacement, $x\,\mathrm{m}$, from a fixed point O on the line is given by $x = 4t^3 - 39t^2 + 120t$.

Find the distance between the two points where P is instantaneously at rest.

6 At time t seconds, where $t \geqslant 0$, the velocity $v\,\mathrm{m\,s^{-1}}$ of a particle moving in a straight line is given by $v = 12 + t - 6t^2$. When $t = 0$, P is at a point O on the line. Find

a the magnitude of the acceleration of P when $v = 0$,

b the distance of P from O when $v = 0$.

7 A particle P is moving on the x-axis. At time t seconds, the velocity of P is $(4t - t^2)\,\mathrm{m\,s^{-1}}$ in the direction of x increasing. At time $t = 0$, P is at the origin O. Find

a the value of x at the instant when $t > 0$ and P is at rest,

b the total distance moved by P in the interval $0 \leqslant t \leqslant 5$.

8 A particle P is moving on the x-axis. At time t seconds, the velocity of P is $(6t^2 - 26t + 15)\,\mathrm{m\,s^{-1}}$ in the direction of x increasing. At time $t = 0$, P is at the origin O. In the subsequent motion P passes through O twice. Find

a the two non-zero values of t when P passes through O,

b the acceleration of P for these two values of t.

9 A particle P of mass 0.4 kg is moving in a straight line under the action of a single variable force $\mathbf{F}$ newtons. At time t seconds (where $t \geqslant 0$) the displacement $x\,\mathrm{m}$ of P from a fixed point O is given by $x = 2t + \dfrac{k}{t + 1}$, where k is a constant. Given that when $t = 0$, the velocity of P is $6\,\mathrm{m\,s^{-1}}$, find

a the value of k,

b the distance of P from O when $t = 0$,

c the magnitude of $\mathbf{F}$ when $t = 3$.

10 A particle P moves along the x-axis. At time t seconds (where $t \geqslant 0$) the velocity of P is $(3t^2 - 12t + 5)\,\mathrm{m\,s^{-1}}$ in the direction of x increasing. When $t = 0$, P is at the origin O. Find

a the velocity of P when its acceleration is zero,

b the values of t when P is again at O,

c the distance travelled by P in the interval $2 \leqslant t \leqslant 3$.

11 A particle P moves in a straight line so that, at time t seconds, its velocity $v\,\mathrm{m\,s^{-1}}$ is given by

$$v = \begin{cases} 4, & 0 \leqslant t \leqslant 2 \\ 5 - \dfrac{4}{t^2}, & t > 2. \end{cases}$$

a Sketch a velocity–time graph to illustrate the motion of P.

b Find the distance moved by P in the interval $0 \leqslant t \leqslant 5$.

12 A particle P moves in a straight line so that, at time t seconds, its acceleration, $a\,\mathrm{m\,s^{-2}}$, is given by

$$a = \begin{cases} 6t - t^2, & 0 \leqslant t \leqslant 2 \\ 8 - t, & t > 2. \end{cases}$$

When $t = 0$ the particle is at rest at a fixed point O on the line. Find

a the speed of P when $t = 2$,

b the speed of P when $t = 4$,

c the distance from O to P when $t = 4$.

1.3 You can use calculus, with vectors, for a particle moving in a plane.

When a particle is moving in a plane you can describe its position $\mathbf{r}$, its velocity $\mathbf{v}$ and its acceleration $\mathbf{a}$ using vectors. The relationships between position (displacement), velocity and acceleration are the same in two dimensions as in one dimension:

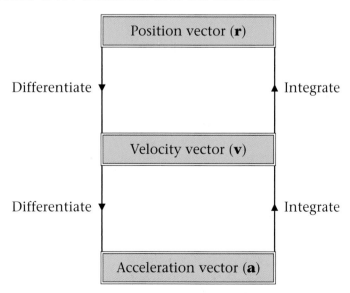

Often a dot is used as a short notation for differentiating with respect to time. One dot is used when you differentiate once with respect to time and two dots are used when you differentiate twice.

■ If $\mathbf{r} = x\mathbf{i} + y\mathbf{j}$, then $\mathbf{v} = \dfrac{d\mathbf{r}}{dt} = \dot{\mathbf{r}} = \dot{x}\mathbf{i} + \dot{y}\mathbf{j}$

and $\mathbf{a} = \dfrac{d\mathbf{v}}{dt} = \dfrac{d^2\mathbf{r}}{dt^2} = \ddot{\mathbf{r}} = \ddot{x}\mathbf{i} + \ddot{y}\mathbf{j}$

$$\dot{x} = \frac{dx}{dt} \quad \text{and} \quad \dot{y} = \frac{dy}{dt}$$

$$\ddot{x} = \frac{d^2x}{dt^2} \quad \text{and} \quad \ddot{y} = \frac{d^2y}{dt^2}$$

■ When you integrate a vector with respect to time, the constant of integration is a vector. Many questions include information which enables you to find this constant vector.

Example 12

A particle P is moving in a plane. At time t seconds, its velocity $\mathbf{v}\,\text{m s}^{-1}$ is given by

$$\mathbf{v} = 3t\mathbf{i} + \tfrac{1}{2}t^2\mathbf{j}$$

When $t = 0$, the position vector of P with respect to a fixed origin O is $(2\mathbf{i} - 3\mathbf{j})\,\text{m}$. Find

a the position vector of P at time t seconds,

b the acceleration of P when $t = 3$.

a $\mathbf{r} = \displaystyle\int \mathbf{v}\,dt = \int \left(3t\mathbf{i} + \tfrac{1}{2}t^2\mathbf{j}\right)dt$

$\qquad = \dfrac{3t^2}{2}\mathbf{i} + \dfrac{t^3}{6}\mathbf{j} + \mathbf{C}$

When $t = 0$, $\mathbf{r} = 2\mathbf{i} - 3\mathbf{j}$

$\qquad 2\mathbf{i} - 3\mathbf{j} = 0\mathbf{i} + 0\mathbf{j} + \mathbf{C}$

$\qquad\qquad \mathbf{C} = 2\mathbf{i} - 3\mathbf{j}$

Hence

$\mathbf{r} = \dfrac{3t^2}{2}\mathbf{i} + \dfrac{t^3}{6}\mathbf{j} + 2\mathbf{i} - 3\mathbf{j} = \left(\dfrac{3t^2}{2} + 2\right)\mathbf{i} + \left(\dfrac{t^3}{6} - 3\right)\mathbf{j}$

The position vector of P at time t seconds is

$\left[\left(\dfrac{3t^2}{2} + 2\right)\mathbf{i} + \left(\dfrac{t^3}{6} - 3\right)\mathbf{j}\right]\text{m}$.

b $\mathbf{v} = 3t\mathbf{i} + \tfrac{1}{2}t^2\mathbf{j}$

$\quad \mathbf{a} = \dot{\mathbf{v}} = 3\mathbf{i} + t\mathbf{j}$

When $t = 3$

$\quad \mathbf{a} = 3\mathbf{i} + 3\mathbf{j}$

When $t = 3$, the acceleration of P is $(3\mathbf{i} + 3\mathbf{j})\,\text{m s}^{-2}$.

You integrate $3t$ and $\tfrac{1}{2}t^2$ in the usual way, using $\displaystyle\int t^n\,dt = \dfrac{t^{n+1}}{n+1}$. You must include the constant of integration, which is a vector, $\mathbf{C}$.

The extra information in the question giving you the position vector when $t = 0$ enables you to find $\mathbf{C}$. You substitute $t = 0$ and $\mathbf{r} = 2\mathbf{i} - 3\mathbf{j}$ into your integrated expression and solve for $\mathbf{C}$.

Collect together the terms in $\mathbf{i}$ and $\mathbf{j}$ to complete your answer.

You differentiate $3t$ and $\tfrac{1}{2}t^2$ in the usual way, using $\dfrac{d}{dt}(t^n) = nt^{n-1}$.

Example 13

A particle P is moving in a plane so that, at time t seconds, its acceleration is $(4\mathbf{i} - 2t\mathbf{j})\,\mathrm{m\,s^{-2}}$. When $t = 3$, the velocity of P is $6\mathbf{i}\,\mathrm{m\,s^{-1}}$ and the position vector of P is $(20\mathbf{i} + 3\mathbf{j})\,\mathrm{m}$ with respect to a fixed origin O. Find

a the angle between the direction of motion of P and $\mathbf{i}$ when $t = 2$,

b the distance of P from O when $t = 0$.

a $\quad \mathbf{v} = \int \mathbf{a}\,dt = \int (4\mathbf{i} - 2t\mathbf{j})\,dt$

$\qquad\qquad = 4t\mathbf{i} - t^2\mathbf{j} + \mathbf{C}$

When $t = 3$, $\mathbf{v} = 6\mathbf{i}$

$\qquad 6\mathbf{i} = 12\mathbf{i} - 9\mathbf{j} + \mathbf{C}$

$\qquad \mathbf{C} = -6\mathbf{i} + 9\mathbf{j}$

Hence

$\qquad \mathbf{v} = 4t\mathbf{i} - t^2\mathbf{j} - 6\mathbf{i} + 9\mathbf{j}$

$\qquad\quad = (4t - 6)\mathbf{i} + (9 - t^2)\mathbf{j}$

When $t = 2$

$\qquad \mathbf{v} = (8 - 6)\mathbf{i} + (9 - 4)\mathbf{j} = 2\mathbf{i} + 5\mathbf{j}$

The angle $\mathbf{v}$ makes with $\mathbf{i}$ is given by

$\qquad \tan\theta = \dfrac{5}{2} \Rightarrow \theta \approx 68.2°$

When $t = 2$, the angle between the direction of motion of P and $\mathbf{i}$ is $68.2°$, to the nearest $0.1°$.

> The direction of motion of P is the direction of the velocity vector of P. Your first step is to find the velocity by integrating the acceleration. You then use the fact that the velocity is $6\mathbf{i}\,\mathrm{m\,s^{-1}}$ when $t = 3$ to find the constant of integration.

> You find the angle the velocity vector makes with $\mathbf{i}$ using trigonometry.

b $\quad \mathbf{r} = \int \mathbf{v}\,dt = \int ((4t - 6)\mathbf{i} + (9 - t^2)\mathbf{j})\,dt$

$\qquad\qquad = (2t^2 - 6t)\mathbf{i} + \left(9t - \dfrac{t^3}{3}\right)\mathbf{j} + \mathbf{D}$

When $t = 3$, $\mathbf{r} = 20\mathbf{i} + 3\mathbf{j}$

$\quad 20\mathbf{i} + 3\mathbf{j} = (18 - 18)\mathbf{i} + (27 - 9)\mathbf{j} + \mathbf{D}$

$\qquad\qquad\quad = 18\mathbf{j} + \mathbf{D}$

$\qquad\qquad \mathbf{D} = 20\mathbf{i} - 15\mathbf{j}$

Hence

$\qquad \mathbf{r} = (2t^2 - 6t)\mathbf{i} + \left(9t - \dfrac{t^3}{3}\right)\mathbf{j} + 20\mathbf{i} - 15\mathbf{j}$

When $t = 0$, $\mathbf{r} = 20\mathbf{i} - 15\mathbf{j}$

$\qquad OP^2 = 20^2 + 15^2 = 625 \Rightarrow OP = \sqrt{625} = 25$

When $t = 0$, the distance of P from O is $25\,\mathrm{m}$.

> You find the position vector by integrating the velocity vector. Remember to include the constant of integration.

> The constant of integration is a vector. This constant is different from the constant in part **a** so you should give it a different symbol.

> You substitute $t = 0$ and find the distance of P from O using Pythagoras' Theorem.

Example 14

A particle P of mass 0.5 kg is moving under the action of a single force $\mathbf{F}$ newtons. At time t seconds, the position vector of P, $\mathbf{r}$ metres, is given by

$$\mathbf{r} = \left(\frac{3t^2}{2} - \frac{t^3}{3}\right)\mathbf{i} + (2t^2 - 8t)\mathbf{j}$$

Find

a the value of t when P is moving parallel to the vector $\mathbf{i}$,

b the magnitude of $\mathbf{F}$ when $t = 3.5$.

a

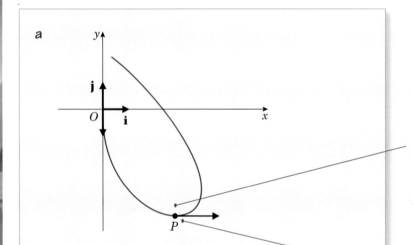

$$\mathbf{r} = \left(\frac{3t^2}{2} - \frac{t^3}{3}\right)\mathbf{i} + (2t^2 - 8t)\mathbf{j}$$

$$\mathbf{v} = \dot{\mathbf{r}} = (3t - t^2)\mathbf{i} + (4t - 8)\mathbf{j}$$

When P is moving parallel to $\mathbf{i}$, the component of $\mathbf{v}$ in the direction of $\mathbf{j}$ is 0.

$$4t - 8 = 0 \Rightarrow t = 2$$

b $\mathbf{a} = \dot{\mathbf{v}} = (3 - 2t)\mathbf{i} + 4\mathbf{j}$

Using Newton's second law

$$\mathbf{F} = m\mathbf{a}$$

$$\mathbf{F} = 0.5((3 - 2t)\mathbf{i} + 4\mathbf{j})$$

When $t = 3.5$

$$\mathbf{F} = 0.5(-4\mathbf{i} + 4\mathbf{j}) = -2\mathbf{i} + 2\mathbf{j}$$

The magnitude of $\mathbf{F}$ is given by

$$|\mathbf{F}|^2 = (-2)^2 + 2^2 = 8 \Rightarrow |\mathbf{F}| = \sqrt{8} = 2\sqrt{2}$$

The magnitude of $\mathbf{F}$ is $2\sqrt{2}$.

The particle starts at the origin and moves along the path shown in this diagram. There is one point where P is moving parallel to $\mathbf{i}$ and you are asked to find the value of t at this point.

When P is moving in the direction of $\mathbf{i}$, its velocity has no component in the direction of $\mathbf{j}$.

You find $\mathbf{F}$ using the law $\mathbf{F} = m\mathbf{a}$. You begin this part by differentiating the velocity to find the acceleration.

Finding the magnitude of vectors is in Chapter 6 of book M1.
If $\mathbf{F} = x\mathbf{i} + y\mathbf{j}$, then $|\mathbf{F}|^2 = x^2 + y^2$.

Example 15

The velocity of a particle P at time t seconds is $((3t^2 - 8)\mathbf{i} + 5\mathbf{j})\,\mathrm{m\,s^{-1}}$. When $t = 0$, the position vector of P with respect to a fixed origin O is $(2\mathbf{i} - 4\mathbf{j})\,\mathrm{m}$.

a Find the position vector of P after t seconds.

A second particle Q moves with constant velocity $(8\mathbf{i} + 4\mathbf{j})\,\mathrm{m\,s^{-1}}$. When $t = 0$, the position vector of Q with respect to the fixed origin O is $2\mathbf{i}\,\mathrm{m}$.

b Prove that P and Q collide.

a Let the position vector of P after t seconds be $\mathbf{p}$ metres.

$$\mathbf{p} = \int \mathbf{v}\,dt = \int ((3t^2 - 8)\mathbf{i} + 5\mathbf{j})\,dt$$

$$\mathbf{p} = (t^3 - 8t)\mathbf{i} + 5t\mathbf{j} + \mathbf{C}$$

When $t = 0$, $\mathbf{p} = 2\mathbf{i} - 4\mathbf{j}$

$$2\mathbf{i} - 4\mathbf{j} = 0\mathbf{i} + 0\mathbf{j} + \mathbf{C} \Rightarrow \mathbf{C} = 2\mathbf{i} - 4\mathbf{j}$$

Hence

$$\mathbf{p} = (t^3 - 8t)\mathbf{i} + 5t\mathbf{j} + 2\mathbf{i} - 4\mathbf{j} = (t^3 - 8t + 2)\mathbf{i} + (5t - 4)\mathbf{j}$$

The position vector of P after t seconds is

$$((t^3 - 8t + 2)\mathbf{i} + (5t - 4)\mathbf{j})\,\mathrm{m}.$$

> There are two position vectors in this question and to write them both as $\mathbf{r}$ m would be confusing. It is sensible to write the position vector of P as $\mathbf{p}$ m and the position vector of Q as $\mathbf{q}$ m.

b Let the position vector of Q after t seconds be $\mathbf{q}$ metres.

$$\mathbf{q} = \int \mathbf{v}\,dt = \int (8\mathbf{i} + 4\mathbf{j})\,dt$$

$$\mathbf{q} = 8t\mathbf{i} + 4t\mathbf{j} + \mathbf{D}$$

When $t = 0$, $\mathbf{q} = 2\mathbf{i}$

$$2\mathbf{i} = 0\mathbf{i} + 0\mathbf{j} + \mathbf{D} \Rightarrow \mathbf{D} = 2\mathbf{i}$$

Hence

$$\mathbf{q} = 8t\mathbf{i} + 4t\mathbf{j} + 2\mathbf{i} = (8t + 2)\mathbf{i} + 4t\mathbf{j}$$

For the $\mathbf{j}$ components of the position vectors to be the same

$$5t - 4 = 4t \Rightarrow t = 4$$

When $t = 4$,

the $\mathbf{i}$ component of $\mathbf{p}$ is $4^3 - 8 \times 4 + 2 = 34$,

the $\mathbf{i}$ component of $\mathbf{q}$ is $8 \times 4 + 2 = 34$.

When $t = 4$

$$\mathbf{p} = \mathbf{q} = 34\mathbf{i} + 16\mathbf{j}$$

Hence P and Q collide.

> Most questions in this section have variable velocity and acceleration. However, you can use the relations between position, velocity and acceleration given in this section if the acceleration or the velocity is constant. If k is a constant, $\int k\,dt = kt$ plus a constant of integration.

> For P and Q to collide, they must be at the same point at the same time. You must find a value of t for which $\mathbf{p}$ and $\mathbf{q}$ are equal. This calculation shows that when $t = 4$, the $\mathbf{i}$ components of $\mathbf{p}$ and $\mathbf{q}$ are both 34 and the $\mathbf{j}$ components of $\mathbf{p}$ and $\mathbf{q}$ are both 16.

Exercise 1C

1 At time t seconds, a particle P has position vector $\mathbf{r}$ m with respect to a fixed origin O, where
$$\mathbf{r} = (3t - 4)\mathbf{i} + (t^3 - 4t)\mathbf{j}.$$
Find
a the velocity of P when $t = 3$,
b the acceleration of P when $t = 3$.

2 A particle P is moving in a plane with velocity $\mathbf{v}$ m s^{-1} at time t seconds where
$$\mathbf{v} = t^2\mathbf{i} + (2t - 3)\mathbf{j}.$$
When $t = 0$, P has position vector $(3\mathbf{i} + 4\mathbf{j})$ m with respect to a fixed origin O. Find
a the acceleration of P at time t seconds,
b the position vector of P when $t = 1$.

3 A particle P starts from rest at a fixed origin O. The acceleration of P at time t seconds (where $t \geqslant 0$) is $(6t^2\mathbf{i} + (8 - 4t^3)\mathbf{j})$ m s^{-2}. Find
a the velocity of P when $t = 2$,
b the position vector of P when $t = 4$.

4 At time t seconds, a particle P has position vector $\mathbf{r}$ m with respect to a fixed origin O, where
$$\mathbf{r} = 4t^2\mathbf{i} + (24t - 3t^2)\mathbf{j}.$$
a Find the speed of P when $t = 2$.
b Show that the acceleration of P is a constant and find the magnitude of this acceleration.

5 A particle P is initially at a fixed origin O. At time $t = 0$, P is projected from O and moves so that, at time t seconds after projection, its position vector $\mathbf{r}$ m relative to O is given by
$$\mathbf{r} = (t^3 - 12t)\mathbf{i} + (4t^2 - 6t)\mathbf{j}, \ t \geqslant 0.$$
Find
a the speed of projection of P,
b the value of t at the instant when P is moving parallel to $\mathbf{j}$,
c the position vector of P at the instant when P is moving parallel to $\mathbf{j}$.

6 At time t seconds, the force $\mathbf{F}$ newtons acting on a particle P, of mass 0.5 kg , is given by
$$\mathbf{F} = 3t\mathbf{i} + (4t - 5)\mathbf{j}.$$
When $t = 1$, the velocity of P is $12\mathbf{i}$ m s^{-1}. Find
a the velocity of P after t seconds,
b the angle the direction of motion of P makes with $\mathbf{i}$ when $t = 5$, giving your answer to the nearest degree.

7 A particle P is moving in a plane with velocity $\mathbf{v}\,\mathrm{m\,s^{-1}}$ at time t seconds where
$$\mathbf{v} = (3t^2 + 2)\mathbf{i} + (6t - 4)\mathbf{j}.$$
When $t = 2$, P has position vector $9\mathbf{j}\,\mathrm{m}$ with respect to a fixed origin O. Find

a the distance of P from O when $t = 0$,

b the acceleration of P at the instant when it is moving parallel to the vector $\mathbf{i}$.

8 At time t seconds, the particle P is moving in a plane with velocity $\mathbf{v}\,\mathrm{m\,s^{-1}}$ and acceleration $\mathbf{a}\,\mathrm{m\,s^{-2}}$, where
$$\mathbf{a} = (2t - 4)\mathbf{i} + 6\mathbf{j}.$$
Given that P is instantaneously at rest when $t = 4$, find

a $\mathbf{v}$ in terms of t,

b the speed of P when $t = 5$.

9 A particle P is moving in a plane. At time t seconds, the position vector of P, $\mathbf{r}\,\mathrm{m}$, is given by
$$\mathbf{r} = (3t^2 - 6t + 4)\mathbf{i} + (t^3 + kt^2)\mathbf{j}, \text{ where } k \text{ is a constant.}$$
When $t = 3$, the speed of P is $12\sqrt{5}\,\mathrm{m\,s^{-1}}$.

a Find the two possible values of k.

b For both of these values of k, find the magnitude of the acceleration of P when $t = 1.5$.

10 At time t seconds (where $t \geqslant 0$), the particle P is moving in a plane with acceleration $\mathbf{a}\,\mathrm{m\,s^{-2}}$, where
$$\mathbf{a} = (5t - 3)\mathbf{i} + (8 - t)\mathbf{j}$$
When $t = 0$, the velocity of P is $(2\mathbf{i} - 5\mathbf{j})\,\mathrm{m\,s^{-1}}$. Find

a the velocity of P after t seconds,

b the value of t for which P is moving parallel to $\mathbf{i} - \mathbf{j}$,

c the speed of P when it is moving parallel to $\mathbf{i} - \mathbf{j}$.

11 At time t seconds (where $t \geqslant 0$), a particle P is moving in a plane with acceleration $(2\mathbf{i} - 2t\mathbf{j})\,\mathrm{m\,s^{-2}}$. When $t = 0$, the velocity of P is $2\mathbf{j}\,\mathrm{m\,s^{-1}}$ and the position vector of P is $6\mathbf{i}\,\mathrm{m}$ with respect to a fixed origin P.

a Find the position vector of P at time t seconds.

At time t seconds (where $t \geqslant 0$), a second particle Q is moving in the plane with velocity $((3t^2 - 4)\mathbf{i} - 2t\mathbf{j})\,\mathrm{m\,s^{-1}}$. The particles collide when $t = 3$.

b Find the position vector of Q at time $t = 0$.

12 A particle P of mass $0.2\,\mathrm{kg}$ is at rest at a fixed origin O. At time t seconds, where $0 \leqslant t \leqslant 3$, a force $(2t\mathbf{i} + 3\mathbf{j})\,\mathrm{N}$ is applied to P.

a Find the position vector of P when $t = 3$.

When $t = 3$, the force acting on P changes to $(6\mathbf{i} + (12 - t^2)\mathbf{j})\,\mathrm{N}$, where $t \geqslant 3$.

b Find the velocity of P when $t = 6$.

Mixed exercise **1D**

Whenever a numerical value of g is required, take $g = 9.8\,\text{m s}^{-2}$.

1 A particle P is projected from a point O on a horizontal plane with speed $42\,\text{m s}^{-1}$ and with angle of elevation $45°$. After projection, the particle moves freely under gravity until it strikes the plane. Find

    a the greatest height above the plane reached by P,

    b the time of flight of P.

2 A stone is thrown horizontally with speed $21\,\text{m s}^{-1}$ from a point P on the edge of a cliff h metres above sea level. The stone lands in the sea at a point Q, where the horizontal distance of Q from the cliff ís $56\,\text{m}$.

    Calculate the value of h.

3 A particle P moves in a horizontal straight line. At time t seconds (where $t \geqslant 0$) the velocity $v\,\text{m s}^{-1}$ of P is given by $v = 15 - 3t$. Find

    a the value of t when P is instantaneously at rest,

    b the distance travelled by P between the time when $t = 0$ and the time when P is instantaneously at rest.

4 A particle P moves along the x-axis so that, at time t seconds, the displacement of P from O is x metres and the velocity of P is $v\,\text{m s}^{-1}$, where

$$v = 6t + \tfrac{1}{2}t^3.$$

    a Find the acceleration of P when $t = 4$.

    b Given also that $x = -5$ when $t = 0$, find the distance OP when $t = 4$.

5 At time t seconds, a particle P has position vector $\mathbf{r}$ m with respect to a fixed origin O, where

$$\mathbf{r} = (3t^2 - 4)\mathbf{i} + (8 - 4t^2)\mathbf{j}.$$

    a Show that the acceleration of P is a constant.

    b Find the magnitude of the acceleration of P and the size of the angle which the acceleration makes with $\mathbf{j}$.

6 At time $t = 0$ a particle P is at rest at a point with position vector $(4\mathbf{i} - 6\mathbf{j})$ m with respect to a fixed origin O. The acceleration of P at time t seconds (where $t \geqslant 0$) is $((4t - 3)\mathbf{i} - 6t^2\mathbf{j})\,\text{m s}^{-2}$. Find

    a the velocity of P when $t = \tfrac{1}{2}$,

    b the position vector of P when $t = 6$.

7 A ball is thrown from a window above a horizontal lawn. The velocity of projection is $15\,\text{m s}^{-1}$ and the angle of elevation is α, where $\tan \alpha = \tfrac{4}{3}$. The ball takes $4\,\text{s}$ to reach the lawn. Find

    a the horizontal distance between the point of projection and the point where the ball hits the lawn,

    b the vertical height above the lawn from which the ball was thrown.

8 A projectile is fired with velocity $40\,\mathrm{m\,s^{-1}}$ at an angle of elevation of $30°$ from a point A on horizontal ground. The projectile moves freely under gravity until it reaches the ground at the point B. Find

a the distance AB,

b the speed of the projectile at the instants when it is $15\,\mathrm{m}$ above the plane.

9 At time t seconds, a particle P has position vector $\mathbf{r}$ m with respect to a fixed origin O, where
$$\mathbf{r} = 2\cos 3t\mathbf{i} - 2\sin 3t\mathbf{j}.$$

a Find the velocity of P when $t = \dfrac{\pi}{6}$.

b Show that the magnitude of the acceleration of P is constant.

10 A particle P of mass $0.2\,\mathrm{kg}$ is moving in a straight line under the action of a single variable force $\mathbf{F}$ newtons. At time t seconds the displacement, s metres, of P from a fixed point A is given by $s = 3t + 4t^2 - \frac{1}{2}t^3$.
Find the magnitude of $\mathbf{F}$ when $t = 4$.

11 At time t seconds (where $t \geqslant 0$) the particle P is moving in a plane with acceleration $\mathbf{a}\,\mathrm{m\,s^{-2}}$, where
$$\mathbf{a} = (8t^3 - 6t)\mathbf{i} + (8t - 3)\mathbf{j}.$$
When $t = 2$, the velocity of P is $(16\mathbf{i} + 3\mathbf{j})\,\mathrm{m\,s^{-1}}$. Find

a the velocity of P after t seconds,

b the value of t when P is moving parallel to $\mathbf{i}$.

12 A particle of mass $0.5\,\mathrm{kg}$ is acted upon by a variable force $\mathbf{F}$. At time t seconds, the velocity $\mathbf{v}\,\mathrm{m\,s^{-1}}$ is given by
$$\mathbf{v} = (4ct - 6)\mathbf{i} + (7 - c)t^2\mathbf{j}, \text{ where } c \text{ is a constant.}$$

a Show that $\mathbf{F} = [2c\mathbf{i} + (7 - c)t\mathbf{j}]\,\mathrm{N}$.

b Given that when $t = 5$ the magnitude of $\mathbf{F}$ is $17\,\mathrm{N}$, find the possible values of c.

13 A ball, attached to the end of an elastic string, is moving in a vertical line. The motion of the ball is modelled as a particle B moving along a vertical axis so that its displacement, x m, from a fixed point O on the line at time t seconds is given by $x = 0.6\cos\left(\dfrac{\pi t}{3}\right)$. Find

a the distance of B from O when $t = \frac{1}{2}$,

b the smallest positive value of t for which B is instantaneously at rest,

c the magnitude of the acceleration of B when $t = 1$. Give your answer to 3 significant figures.

14 A light spot S moves along a straight line on a screen. At time $t = 0$, S is at a point O. At time t seconds (where $t \geqslant 0$) the distance, x cm, of S from O is given by $x = 4t\,\mathrm{e}^{-0.5t}$. Find

a the acceleration of S when $t = \ln 4$,

b the greatest distance of S from O.

15 A particle P is projected with velocity $(3u\mathbf{i} + 4u\mathbf{j})\,\text{m s}^{-1}$ from a fixed point O on a horizontal plane. Given that P strikes the plane at a point $750\,\text{m}$ from O,

   a show that $u = 17.5$,

   b calculate the greatest height above the plane reached by P,

   c find the angle the direction of motion of P makes with $\mathbf{i}$ when $t = 5$.

16 A particle P is projected from a point on a horizontal plane with speed u at an angle of elevation θ.

   a Show that the range of the projectile is $\dfrac{u^2 \sin 2\theta}{g}$.

   b Hence find, as θ varies, the maximum range of the projectile.

   c Given that the range of the projectile is $\dfrac{2u^2}{3g}$, find the two possible value of θ. Give your answers to $0.1°$.

17

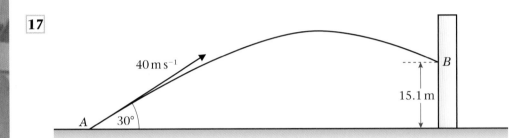

A golf ball is driven from a point A with a speed of $40\,\text{m s}^{-1}$ at an angle of elevation of $30°$. On its downward flight, the ball hits an advertising hoarding at a height $15.1\,\text{m}$ above the level of A, as shown in the diagram above. Find

   a the time taken by the ball to reach its greatest height above A,

   b the time taken by the ball to travel from A to B,

   c the speed with which the ball hits the hoarding.

18 A particle P passes through a point O and moves in a straight line. The displacement, s metres, of P from O, t seconds after passing through O is given by

$$s = -t^3 + 11t^2 - 24t$$

   a Find an expression for the velocity, $v\,\text{m s}^{-1}$, of P at time t seconds.

   b Calculate the values of t at which P is instantaneously at rest.

   c Find the value of t at which the acceleration is zero.

   d Sketch a velocity-time graph to illustrate the motion of P in the interval $0 \leqslant t \leqslant 6$, showing on your sketch the coordinates of the points at which the graph crosses the axes.

   e Calculate the values of t in the interval $0 \leqslant t \leqslant 6$ between which the speed of P is greater than $16\,\text{m s}^{-1}$. **E**

19 A point P moves in a straight line so that, at time t seconds, its displacement from a fixed point O on the line is given by

$$s = \begin{cases} 4t^2, & 0 \leqslant t \leqslant 3 \\ 24t - 36, & 3 < t \leqslant 6 \\ -252 + 96t - 6t^2, & t > 6. \end{cases}$$

Find

a the velocity of P when $t = 4$,

b the velocity of P when $t = 10$,

c the greatest positive displacement of P from O,

d the values of s when the speed of P is $18 \, \text{m s}^{-1}$.

20 The position vector of a particle P, with respect to a fixed origin O, at time t seconds (where $t \geqslant 0$) is $\left[\left(6t - \frac{1}{2}t^3 \right)\mathbf{i} + (3t^2 - 8t)\mathbf{j} \right]$ m. At time t seconds, the velocity of a second particle Q, moving in the same plane as P, is $(-8\mathbf{i} + 3t\mathbf{j}) \, \text{m s}^{-1}$.

a Find the value of t at the instant when the direction of motion of P is perpendicular to the x-axis.

b Given that P and Q collide when $t = 4$, find the position vector of Q with respect to O when $t = 0$.

Summary of key points

1 The vertical motion of the projectile is motion with constant acceleration $g = 9.8 \, \text{m s}^{-2}$. You can use the constant acceleration formulae from M1.

2 The horizontal motion of a projectile is motion with constant speed. You can use the formula **distance = speed × time**.

3 Velocity is the rate of change of displacement with time. To find the velocity from the displacement, you differentiate with respect to time.

Using symbols:

$$v = \frac{dx}{dt}$$

$$a = \frac{dv}{dt} = \frac{d^2x}{dt^2}$$

4 Acceleration is the rate of change of velocity with time. To find the acceleration from the velocity you differentiate with respect to time.

$$v = \int a \, dt$$

$$x = \int v \, dt$$

5 To obtain the velocity from the acceleration, you integrate with respect to time.

6 To obtain the displacement from the velocity, you integrate with respect to time.

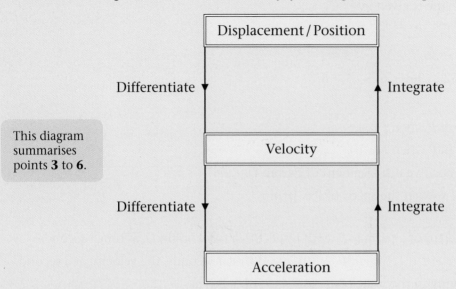

This diagram summarises points **3** to **6**.

7 Often a dot is used as a short notation for differentiating with respect to time. One dot is used when you differentiate once with respect to time and two dots are used when you differentiate twice.

8 If $\mathbf{r} = x\mathbf{i} + y\mathbf{j}$, then $\mathbf{v} = \dfrac{\mathrm{d}\mathbf{r}}{\mathrm{d}t} = \dot{\mathbf{r}} = \dot{x}\mathbf{i} + \dot{y}\mathbf{j}$

and $\mathbf{a} = \dfrac{\mathrm{d}\mathbf{v}}{\mathrm{d}t} = \dfrac{\mathrm{d}^2\mathbf{r}}{\mathrm{d}t^2} = \ddot{\mathbf{r}} = \ddot{x}\mathbf{i} + \ddot{y}\mathbf{j}$

9 When you integrate a vector with respect to time, the constant of integration is a vector.

$$\mathbf{v} = \int \mathbf{a}\,\mathrm{d}t$$

and $\mathbf{r} = \int \mathbf{v}\,\mathrm{d}t$

After completing this chapter you should be able to:

1 find the centre of mass of a system of particles distributed in one dimension

2 find the centre of mass of a system of particles distributed in two dimensions

3 use knowledge of standard results to find the centre of mass of a plane figure

4 consider the equilibrium of a lamina which is suspended from a fixed point or placed on an inclined plane.

Centres of mass

In M1 you learnt how to find the moment of a force and how to solve simple problems involving a number of forces acting on a 'large' body (a body which cannot be modelled as a particle).

In this chapter you will use these ideas to find the position of the **centre of mass** of a large body. This is the point at which the whole mass of the body can be considered to be concentrated.

This tightrope-walker is balanced on a high-wire. He is in equilibrium. His centre of mass is positioned directly above the wire.

2.1 You can use $\sum m_i x_i = \bar{x} \sum m_i$ to find the centre of mass of a set of particles arranged along a straight line.

Example **1**

Find the centre of mass of three particles of masses 2 kg, 5 kg and 3 kg which lie on the x-axis at the points $(3, 0)$, $(4, 0)$ and $(6, 0)$ respectively.

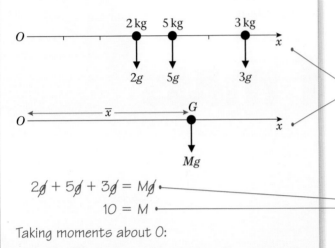

Draw two diagrams, the first showing the weights of the three particles, the second showing the total weight Mg acting at the centre of mass G.

$$2\not{g} + 5\not{g} + 3\not{g} = M\not{g}$$
$$10 = M$$

Compare the two systems vertically. Note that the g cancels.

Taking moments about O:

$$(2\not{g} \times 3) + (5\not{g} \times 4) + (3\not{g} \times 6) = M\not{g} \times \bar{x}$$
$$(2 \times 3) + (5 \times 4) + (3 \times 6) = 10\bar{x}$$

Equate the moments of the two systems about O.
Substitute for M.

$$6 + 20 + 18 = 10\bar{x}$$
$$4.4 = \bar{x}$$

The centre of mass is $(4.4, 0)$

Example **2**

A system of n particles, with masses $m_1, m_2, ..., m_n$ are placed along the x-axis at the points $(x_1, 0), (x_2, 0), ..., (x_n, 0)$ respectively. Find the centre of mass of the system.

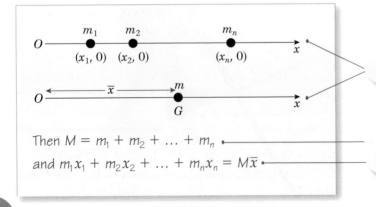

Draw two diagrams.

Then $M = m_1 + m_2 + ... + m_n$

and $m_1 x_1 + m_2 x_2 + ... + m_n x_n = M\bar{x}$

There is no need to include g as it cancels.

I.e. $m_1x_1 + m_2x_2 + \ldots + m_nx_n = (m_1 + m_2 + \ldots + m_n)\bar{x}$

or $\displaystyle\sum_{i=1}^{n} m_ix_i = \bar{x}\sum_{i=1}^{n} m_i$ ──────── This is a key result and should be learned.

So $\bar{x} = \dfrac{\displaystyle\sum_{i=1}^{n} m_ix_i}{\displaystyle\sum_{i=1}^{n} m_i}$ ──────── Note that this result holds for positive and negative coordinates.

■ If a system of n particles with masses m_1, m_2, ..., m_n are placed along the y-axis at the points $(0, y_1)$, $(0, y_2)$, ..., $(0, y_n)$ respectively, then

$$\sum_{i=1}^{n} m_iy_i = \bar{y}\sum_{i=1}^{n} m_i$$

where $(0, \bar{y})$ is the position of the centre of mass of the system.

Exercise 2A

1 Find the position of the centre of mass of four particles of masses 1 kg, 4 kg, 3 kg and 2 kg placed on the x-axis at the points $(6, 0)$, $(3, 0)$ $(2, 0)$ and $(4, 0)$ respectively.

2 Three masses 1 kg, 2 kg and 3 kg, are placed at the points with coordinates $(0, 2)$, $(0, 5)$ and $(0, 1)$ respectively. Find the coordinates of G, the centre of mass of the three masses.

3 Three particles of mass 2 kg, 3 kg and 5 kg, are placed at the points $(-1, 0)$, $(-4, 0)$ and $(5, 0)$ respectively. Find the coordinates of the centre of mass of the three particles.

4 A light rod PQ of length 4 m has particles of mass 1 kg, 2 kg and 3 kg attached to it at the points P, Q and R respectively, where $PR = 2$ m. The centre of mass of the loaded rod is at the point G. Find the distance PG.

5 Three particles of mass 5 kg, 3 kg and m kg lie on the y-axis at the points $(0, 4)$, $(0, 2)$ and $(0, 5)$ respectively. The centre of mass of the system is at the point $(0, 4)$. Find the value of m.

6 A light rod PQ of length 2 m has particles of masses 0.4 kg and 0.6 kg fixed to it at the points P and R respectively, where $PR = 0.5$ m. Find the mass of the particle which must be fixed at Q so that the centre of mass of the loaded rod is at its mid-point.

7 The centre of mass of four particles of masses $2m$, $3m$, $7m$ and $8m$, which are positioned at the points $(0, a)$, $(0, 2)$, $(0, -1)$ and $(0, 1)$ respectively, is the point G. Given that the coordinates of G are $(0, 1)$, find the value of a.

8 Particles of mass 3 kg, 2 kg and 1 kg lie on the y-axis at the points with coordinates $(0, -2)$, $(0, 7)$ and $(0, 4)$ respectively. Another particle of mass 6 kg is added to the system so that the centre of mass of all four particles is at the origin. Find the position of this particle.

9 Three particles A, B and C are placed along the x-axis. Particle A has mass 5 kg and is at the point $(2, 0)$. Particle B has mass m_1 kg and is at the point $(3, 0)$ and particle C has mass m_2 kg and is at the point $(-2, 0)$. The centre of mass of the three particles is at the point G $(1, 0)$. Given that the total mass of the three particles is 10 kg, find the values of m_1 and m_2.

10 Three particles A, B and C have masses 4 kg, 1 kg and 5 kg respectively. The particles are placed on the line with equation $3y - 4x = 0$. Particle A is at the origin, particle B is at the point $(3, 4)$ and particle C is at the point $(9, 12)$. Find the coordinates of the centre of mass of the three particles.

2.2 You can use $\sum m_i x_i = \bar{x} \sum m_i$ and $\sum m_i y_i = \bar{y} \sum m_i$ to find the centre of mass of a set of particles arranged in a plane.

In this section you will extend the ideas from section 2.1 to find the position of the centre of mass of a system of particles which are arranged in a plane.

Example 3

Find the coordinates of the centre of mass of the following system of particles:
2 kg at $(1, 2)$; 3 kg at $(3, 1)$; 5 kg at $(4, 3)$.

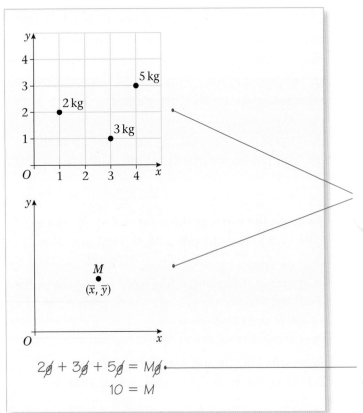

Draw two diagrams, the first showing the three particles, the second showing the total mass M placed at the centre of mass $(\bar{x}, \bar{y})$.

$$2\cancel{g} + 3\cancel{g} + 5\cancel{g} = M\cancel{g}$$
$$10 = M$$

Equate the total weights.
Note that g cancels.

Method 1

Taking moments about the y-axis:

$$(2g \times 1) + (3g \times 3) + (5g \times 4) = Mg\bar{x}$$

Equate the moments of the systems about the y-axis.

$$(2 \times 1) + (3 \times 3) + (5 \times 4) = (2 + 3 + 5)\bar{x}$$

Substitute for M.

$$2 + 9 + 20 = 10\bar{x}$$
$$3.1 = \bar{x}$$

Taking moments about the x-axis:

$$(2g \times 2) + (3g \times 1) + (5g \times 3) = (2 + 3 + 5)g\bar{y}$$

Equate the moments of the systems about the x-axis.

$$2.2 = \bar{y}$$

The centre of mass is (3.1, 2.2).

Method 2

You can reduce the working by using position vectors.

$$2\binom{1}{2} + 3\binom{3}{1} + 5\binom{4}{3} = (2 + 3 + 5)\binom{\bar{x}}{\bar{y}}$$

$$\binom{2}{4} + \binom{9}{3} + \binom{20}{15} = \binom{10\bar{x}}{10\bar{y}}$$

The top line is
$$\sum m_i x_i = \bar{x}\sum m_i$$
and the bottom line is
$$\sum m_i y_i = \bar{y}\sum m_i$$

$$\binom{31}{22} = \binom{10\bar{x}}{10\bar{y}}$$

Simplify.

$$\binom{3.1}{2.2} = \binom{\bar{x}}{\bar{y}}$$

The centre of mass is at (3.1, 2.2).

Divide both sides by 10.

■ In general, if a system consists of n particles: mass m_1 with position vector $\mathbf{r}_1$, mass m_2 with position vector $\mathbf{r}_2$, ... mass m_n with position vector $\mathbf{r}_n$, then

$$\sum m_i \mathbf{r}_i = \bar{\mathbf{r}}\sum m_i$$

where $\bar{\mathbf{r}}$ is the position vector of the centre of mass of the system.

The position vector of a point can be written in terms of $\mathbf{i}$ and $\mathbf{j}$ or as a column vector. For example, the position vector of the point (3, 4) is $3\mathbf{i} + 4\mathbf{j}$ or $\binom{3}{4}$.

Example 4

Find the coordinates of the centre of mass of the following system of particles:

4 kg at $(-1, 3)$; 2 kg at $(-2, -4)$; 8 kg at $(4, 0)$; 6 kg at $(1, -3)$.

$$4\binom{-1}{3} + 2\binom{-2}{-4} + 8\binom{4}{0} + 6\binom{1}{-3} = (4 + 2 + 8 + 6)\binom{\overline{x}}{\overline{y}}$$

The result applies with positive or negative coordinates.

$$\binom{-4}{12} + \binom{-4}{-8} + \binom{32}{0} + \binom{6}{-18} = 20\binom{\overline{x}}{\overline{y}}$$

$$\binom{30}{-14} = 20\binom{\overline{x}}{\overline{y}}$$

Simplify the LHS.

$$\binom{1.5}{-0.7} = \binom{\overline{x}}{\overline{y}}$$

Centre of mass is $(1.5, -0.7)$

■ **If a question does not specify axes or coordinates you will need to choose your own axes and origin.**

Example 5

A light rectangular plate $ABCD$ has $AB = 20$ cm and $AD = 50$ cm. Particles of mass 2 kg, 3 kg, 5 kg and 5 kg are attached to the plate at the points A, B, C and D respectively.
Find the distance of the centre of mass of the loaded plate from:

a AD,

b AB.

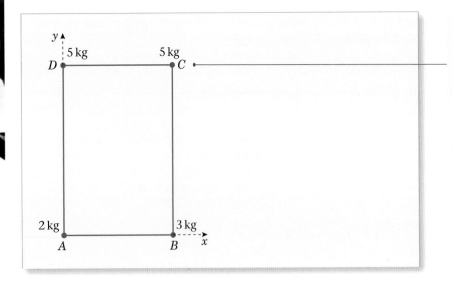

First draw a diagram. Choose point A as the origin and put it in the bottom left-hand corner of your diagram. Using AB and AD as 'axes' A is $(0, 0)$, B is $(20, 0)$, C is $(20, 50)$ and D is $(0, 50)$.

$$2\begin{pmatrix}0\\0\end{pmatrix} + 3\begin{pmatrix}20\\0\end{pmatrix} + 5\begin{pmatrix}20\\50\end{pmatrix} + 5\begin{pmatrix}0\\50\end{pmatrix} = (2 + 3 + 5 + 5)\begin{pmatrix}\bar{x}\\\bar{y}\end{pmatrix}$$

Using $\sum m_i r_i = \bar{r}\sum m_i$

$$\begin{pmatrix}0\\0\end{pmatrix} + \begin{pmatrix}60\\0\end{pmatrix} + \begin{pmatrix}100\\250\end{pmatrix} + \begin{pmatrix}0\\250\end{pmatrix} = 15\begin{pmatrix}\bar{x}\\\bar{y}\end{pmatrix}$$

$$\begin{pmatrix}160\\500\end{pmatrix} = 15\begin{pmatrix}\bar{x}\\\bar{y}\end{pmatrix}$$

Simplify LHS.

$$\begin{pmatrix}\frac{32}{3}\\\frac{100}{3}\end{pmatrix} = \begin{pmatrix}\bar{x}\\\bar{y}\end{pmatrix}$$

Divide both sides by 15.

a $10\frac{2}{3}$ cm

b $33\frac{1}{3}$ cm

Check that your answers are sensible in the context of the question.

Example 6

Particles of mass 4 kg, 3 kg, 2 kg and 1 kg are placed at the points (x, y), $(3, 2)$ $(1, -5)$ and $(6, 0)$ respectively. Given that the centre of mass of the four particles is at the point $(2.5, -2)$, find the values of x and y.

$$4\begin{pmatrix}x\\y\end{pmatrix} + 3\begin{pmatrix}3\\2\end{pmatrix} + 2\begin{pmatrix}1\\-5\end{pmatrix} + 1\begin{pmatrix}6\\0\end{pmatrix} = (4 + 3 + 2 + 1)\begin{pmatrix}2.5\\-2\end{pmatrix}$$

Using $\sum m_i r_i = \bar{r}\sum m_i$

$$\begin{pmatrix}4x\\4y\end{pmatrix} + \begin{pmatrix}9\\6\end{pmatrix} + \begin{pmatrix}2\\-10\end{pmatrix} + \begin{pmatrix}6\\0\end{pmatrix} = \begin{pmatrix}25\\-20\end{pmatrix}$$

Simplify both sides.

$$\begin{pmatrix}4x + 17\\4y - 4\end{pmatrix} = \begin{pmatrix}25\\-20\end{pmatrix}$$

Add the vectors together.

$$4x + 17 = 25$$
$$4y - 4 = -20$$

Equate the **i** and **j** components.

$$4x = 8$$
$$4y = -16$$

Solve the two equations for x and y.

$$x = 2, y = -4$$

Example 7

Three particles of mass 2 kg, 1 kg and m kg are situated at the points $(-1, 3)$, $(2, 9)$ and $(2, -1)$ respectively. Given that the centre of mass of the three particles is at the point $(1, \bar{y})$, find

a the value of m,

b the value of $\bar{y}$.

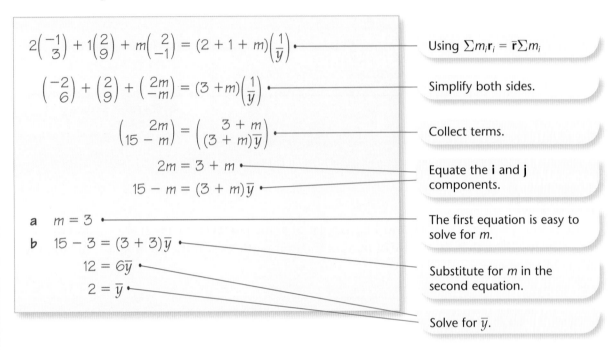

$2\begin{pmatrix} -1 \\ 3 \end{pmatrix} + 1\begin{pmatrix} 2 \\ 9 \end{pmatrix} + m\begin{pmatrix} 2 \\ -1 \end{pmatrix} = (2 + 1 + m)\begin{pmatrix} 1 \\ \bar{y} \end{pmatrix}$	Using $\sum m_i \mathbf{r}_i = \bar{\mathbf{r}} \sum m_i$
$\begin{pmatrix} -2 \\ 6 \end{pmatrix} + \begin{pmatrix} 2 \\ 9 \end{pmatrix} + \begin{pmatrix} 2m \\ -m \end{pmatrix} = (3 + m)\begin{pmatrix} 1 \\ \bar{y} \end{pmatrix}$	Simplify both sides.
$\begin{pmatrix} 2m \\ 15 - m \end{pmatrix} = \begin{pmatrix} 3 + m \\ (3 + m)\bar{y} \end{pmatrix}$	Collect terms.
$2m = 3 + m$	Equate the **i** and **j** components.
$15 - m = (3 + m)\bar{y}$	
a $m = 3$	The first equation is easy to solve for m.
b $15 - 3 = (3 + 3)\bar{y}$	Substitute for m in the second equation.
$12 = 6\bar{y}$	
$2 = \bar{y}$	Solve for $\bar{y}$.

Exercise 2B

1 Two particles of equal mass are placed at the points $(1, -3)$ and $(5, 7)$. Find the centre of mass of the particles.

2 Four particles of equal mass are situated at the points $(2, 0)$, $(-1, 3)$, $(2, -4)$ and $(-1, -2)$. Find the coordinates of the centre of mass of the particles.

3 A system of three particles consists of 10 kg placed at $(2, 3)$, 15 kg placed at $(4, 2)$ and 25 kg placed at $(6, 6)$. Find the coordinates of the centre of mass of the system.

4 Find the position vector of the centre of mass of three particles of masses 0.5 kg, 1.5 kg and 2 kg which are situated at the points with position vectors $(6\mathbf{i} - 3\mathbf{j})$, $(2\mathbf{i} + 5\mathbf{j})$ and $(3\mathbf{i} + 2\mathbf{j})$ respectively.

5 Particles of masses m, $2m$, $5m$ and $2m$ are situated at $(-1, -1)$, $(3, 2)$, $(4, -2)$ and $(-2, 5)$ respectively. Find the coordinates of the centre of mass of the particles.

6 A light rectangular metal plate $PQRS$ has $PQ = 4$ cm and $PS = 2$ cm. Particles of masses 3 kg, 5 kg, 1 kg and 7 kg are attached respectively to the corners P, Q, R and S of the plate. Find the distance of the centre of mass of the loaded plate from

    a the side PQ,   **b** the side PS.

7 Three particles of masses 1 kg, 2 kg and 3 kg are positioned at the points (1, 0), (4, 3) and (*p*, *q*) respectively. Given that the centre of mass of the particles is at the point (2, 0), find the values of *p* and *q*.

8 A system consists of three particles with masses 3*m*, 4*m* and 5*m*. The particles are situated at the points with coordinates (−3, −4), (0.5, 4) and (0, −5) respectively. Find the coordinates of the position of a fourth particle of mass 7*m*, given that the centre of mass of all four particles is at the origin.

9 A light rectangular piece of card *ABCD* has *AB* = 6 cm and *AD* = 4 cm. Four particles of mass 200 g, 300 g, 600 g and 100 g are fixed to the rectangle at the mid-points of the sides *AB*, *BC*, *CA* and *AD* respectively. Find the distance of the centre of mass of the loaded rectangle from

  a the side *AB*,

  b the side *AD*.

10 A light rectangular piece of card *ABCD* has *AB* = 8 cm and *AD* = 6 cm. Three particles of mass 3 g, 2 g and 2 g are attached to the rectangle at the points *A*, *B* and *C* respectively.

  a Find the mass of a particle which must be placed at the point *D* for the centre of mass of the whole system of four particles to lie 3 cm from the line *AB*.

  b With this fourth particle in place, find the distance of the centre of mass of the system from the side *AD*.

2.3 You can find the positions of the centres of mass of standard uniform plane laminas, including a rectangle, a triangle and a semicircle.

■ An object which has one dimension (its thickness) very small compared with the other two (its length and width) is modelled as a **lamina**. This means that it is regarded as being two-dimensional with area rather than volume.

> For example, a sheet of paper or a piece of card could be modelled as a lamina.

A lamina is **uniform** if its mass is evenly spread throughout its area. If a uniform lamina has an axis of symmetry then its centre of mass must lie on the axis of symmetry. If the lamina has more than one axis of symmetry then it follows that the centre of mass must be at the point of intersection of the axes of symmetry.

Uniform circular disc
Since every diameter of the disc is a line of symmetry the centre of mass of the disc is at their intersection. This is the centre of the disc.

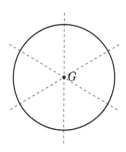

Uniform rectangular lamina

A uniform rectangular lamina has two lines of symmetry,
each one joining the mid-points of a pair of opposite sides.
The centre of mass is at the point where the two lines meet.

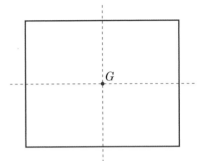

Uniform triangular lamina

A uniform triangular lamina only has axes of symmetry if it is either
equilateral or isosceles.

A uniform equilateral triangle has three axes of symmetry, each one
joining a vertex to the mid-point of the opposite side. These three
lines are called the **medians** of the triangle.

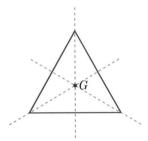

■ **The centre of mass of any uniform
triangular lamina is at the
intersection of the medians.
This point is called the centroid
of the triangle.**

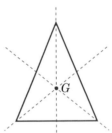

Note that the medians are not axes of
symmetry of the triangle unless the
triangle is equilateral (in which case all
three medians are axes of symmetry) or
isosceles (in which case one median is also
an axis of symmetry).

It can be proved that the
centroid G (and therefore the
centre of mass) of any triangle
is two thirds of the way down
each median from each vertex:

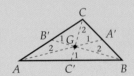

where A' is the mid-point of
BC, B' is the mid-point of CA
and C' is the mid-point of AB

i.e. $\dfrac{AG}{GA'} = \dfrac{BG}{GB'} = \dfrac{CG}{GC'} = \dfrac{2}{1}$

■ **If the coordinates of the three vertices of a uniform triangular lamina are (x_1, y_1), (x_2, y_2)
and (x_3, y_3) then the coordinates of the centre of mass are given by taking the average
(mean) of the coordinates of the vertices:**

G is the point $\left(\dfrac{x_1 + x_2 + x_3}{3}, \dfrac{y_1 + y_2 + y_3}{3} \right)$

The centre of mass of the following plane shape can be found
using methods described in book M3. However, you may
be required to *use* this result in the M2 examination.
It can be found in the formula booklet provided by
Edexcel.

This is the two-dimensional
version of a similar result
for a uniform rod: if the
ends of the rod are (x_1, y_1)
and (x_2, y_2) then its centre
of mass is its mid-point,
the point $\left(\dfrac{x_1 + x_2}{2}, \dfrac{y_1 + y_2}{2} \right)$.

Uniform sector of a circle

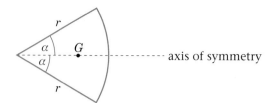

A uniform sector of a circle of radius r and centre angle 2α, where α is measured in radians, has its centre of mass on

the axis of symmetry at a distance $\dfrac{2r \sin \alpha}{3\alpha}$ from the centre.

> This result can be found in the formula booklet provided by Edexcel. You may need to use it in your M2 examination, though it will not be proved until M3.

Example 8

A uniform triangular lamina has vertices $A(1, 4)$, $B(3, 2)$ and $C(5, 3)$. Find the coordinates of its centre of mass.

G is the point $\left(\dfrac{1 + 3 + 5}{3}, \dfrac{4 + 2 + 3}{3}\right) = (3, 3)$

> Find the mean of the vertices of the triangle.

Example 9

Find the centre of mass of the uniform triangular lamina shown:

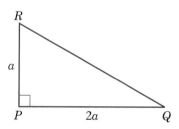

Taking P as the origin and PQ and PR as axes:

P is $(0, 0)$; Q is $(2a, 0)$; R is $(0, a)$

G is the point $\left(\dfrac{0 + 2a + 0}{3}, \dfrac{0 + 0 + a}{3}\right)$

$= \left(\dfrac{2a}{3}, \dfrac{a}{3}\right)$

The centre of mass is $\dfrac{2a}{3}$ from PR and $\dfrac{a}{3}$ from PQ.

> Here we need to choose our own axes and origin.

> Write down the coordinates of each of the three vertices.

> Find the mean of the three vertices.

> When you choose your own axes you must not leave your answer in coordinate form.

Example 10

A light triangular piece of card ABC has vertices $A(2, 1)$, $B(4, 5)$ and $C(9, 3)$. A particle of mass m is placed at each of the three vertices. Find their centre of mass.

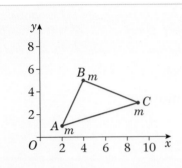

$$m\begin{pmatrix}2\\1\end{pmatrix} + m\begin{pmatrix}4\\5\end{pmatrix} + m\begin{pmatrix}9\\3\end{pmatrix} = (m + m + m)\begin{pmatrix}\bar{x}\\\bar{y}\end{pmatrix}$$

Since the triangle is light, you simply find the centre of mass of the three particles.

$$\begin{pmatrix}2 + 4 + 9\\1 + 5 + 3\end{pmatrix} = 3\begin{pmatrix}\bar{x}\\\bar{y}\end{pmatrix}$$

Use the same approach as in section 2.2

Cancel m.

$$\begin{pmatrix}\dfrac{2 + 4 + 9}{3}\\[2mm]\dfrac{1 + 5 + 3}{3}\end{pmatrix} = \begin{pmatrix}\bar{x}\\\bar{y}\end{pmatrix}$$

Notice that we have found the mean of the vertices of the triangle ABC.

$$\begin{pmatrix}5\\3\end{pmatrix} = \begin{pmatrix}\bar{x}\\\bar{y}\end{pmatrix}$$

Centre of mass is at the point $(5, 3)$

This would also be the centre of mass of a uniform triangular lamina with vertices at A, B and C.

Example 11

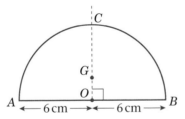

The diagram shows a uniform semi-circular lamina of radius 6 cm with centre O. Find the centre of mass of the lamina.

The centre of mass must lie on the line through O which is perpendicular to AB.

This is the axis of symmetry of the lamina.

Let $OG = \bar{y}$. Then,

$$\bar{y} = \frac{2 \times 6 \times \sin\dfrac{\pi}{2}}{\dfrac{3\pi}{2}}$$

We use the result for a sector which is in the formula booklet with $r = 6$ and $\alpha = \dfrac{\pi}{2}$.

$$= \frac{12 \times 1}{\frac{3\pi}{2}}$$ •————————————— $\sin \frac{\pi}{2} = 1$.
You must give the angle in radians for this formula.

$$= 12 \times \frac{2}{3\pi}$$

$$= \frac{8}{\pi}$$ •————————————— Simplify.

The centre of mass of the lamina is on the line OC at a distance $\frac{8}{\pi}$ cm from O.

Exercise 2C

1 Find the centre of mass of a uniform triangular lamina whose vertices are

a (1, 2), (2, 6) and (3, 1)   **b** (−1, 4), (3, 5) and (7, 3)

c (−3, 2), (4, 0) and (0, 1)   **d** (a, a), ($3a$, $2a$) and ($4a$, $6a$).

2 Find the position of the centre of mass of a uniform semi-circular lamina of radius 4 cm and centre O.

3 The centre of mass of a uniform triangular lamina ABC is at the point (2, a). Given that A is the point (4, 3), B is the point (b, 1) and C is the point (−1, 5), find the values of a and b.

4 Find the position of the centre of mass of the following uniform triangular laminas:

a
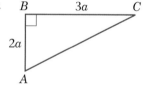

b
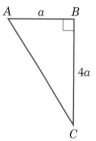

c
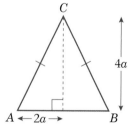

d
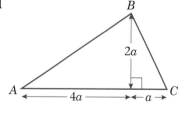

e
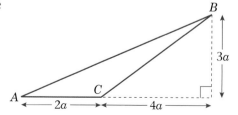

f
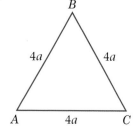

5 A uniform triangular lamina is isosceles and has the line $y = 4$ as its axis of symmetry. One of the vertices of the triangle is the point $(2, 1)$. Given that the x-coordinate of the centre of mass of the lamina is -3, find the coordinates of the other two vertices.

2.4 **You can use $\sum m_i x_i = \bar{x} \sum m_i$ and $\sum m_i y_i = \bar{y} \sum m_i$ together with knowledge of the position of the centre of mass of standard shapes, to find the centre of mass of a composite lamina.**

A uniform plane figure may consist of two or more of the standard uniform shapes mentioned in the previous section, which have been joined together to form a composite plane figure.

Example 12

A uniform lamina consists of a rectangle $PQRS$ joined to an isosceles triangle QRT, as shown in the diagram.

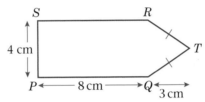

Find the distance of the centre of mass of the lamina from:

a PQ,

b PS.

Let the mass per unit area be m kg per cm².   *Since the lamina is uniform, the mass per unit area will be a constant.*

Split the lamina along the line QR:

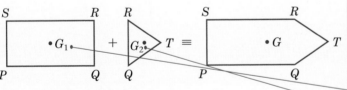

You must always split the lamina up into standard shapes.
G_1 is the centre of mass of the rectangle.
G_2 is the centre of mass of the triangle.

Area of $PQRS = 8 \times 4 = 32$ cm²

So, mass of $PQRS = 32m$ kg

Similarly, mass of $QRT = \frac{1}{2} \times 4 \times 3 \times m$ kg   *Area of △ is $\frac{1}{2} \times$ base $\times$ height.*

                   $= 6m$ kg

So, total mass of the lamina $= 32m + 6m = 38m$ kg

Take P as the origin, and axes along PQ and PS.   *It's usually a good idea to take the origin at the bottom left-hand corner of your diagram.*

The centre of mass of $PQRS$ is at the point $(4, 2)$.   *This is the centre of the rectangle, G_1.*

The coordinates of Q are $(8, 0)$

The coordinates of R are $(8, 4)$

The coordinates of T are $(11, 2)$

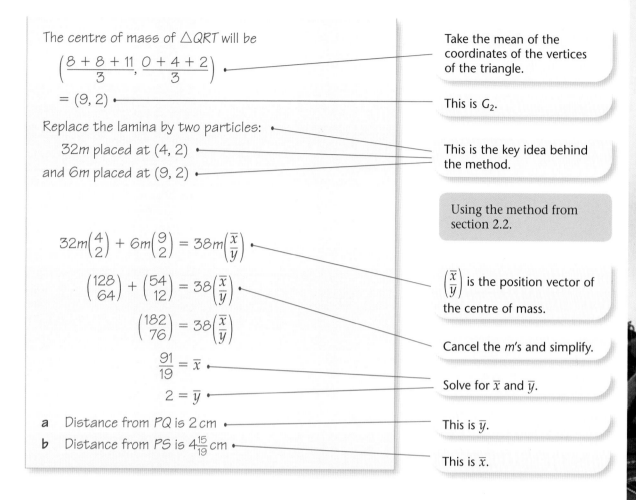

The centre of mass of $\triangle QRT$ will be

$$\left(\frac{8 + 8 + 11}{3}, \frac{0 + 4 + 2}{3}\right)$$

Take the mean of the coordinates of the vertices of the triangle.

$$= (9, 2)$$

This is G_2.

Replace the lamina by two particles:

$32m$ placed at $(4, 2)$

and $6m$ placed at $(9, 2)$

This is the key idea behind the method.

Using the method from section 2.2.

$$32m\binom{4}{2} + 6m\binom{9}{2} = 38m\binom{\bar{x}}{\bar{y}}$$

$\binom{\bar{x}}{\bar{y}}$ is the position vector of the centre of mass.

$$\binom{128}{64} + \binom{54}{12} = 38\binom{\bar{x}}{\bar{y}}$$

$$\binom{182}{76} = 38\binom{\bar{x}}{\bar{y}}$$

Cancel the m's and simplify.

$$\frac{91}{19} = \bar{x}$$

$$2 = \bar{y}$$

Solve for $\bar{x}$ and $\bar{y}$.

a Distance from PQ is $2\,\text{cm}$

This is $\bar{y}$.

b Distance from PS is $4\frac{15}{19}\,\text{cm}$

This is $\bar{x}$.

Note that you could have got the answer to **a** using the fact that the lamina has an axis of symmetry. You should always use this as it will considerably reduce the amount of working required.

■ **The centre of mass of a uniform plane lamina will always lie on an axis of symmetry.**

Example 13

The diagram shows a uniform lamina.

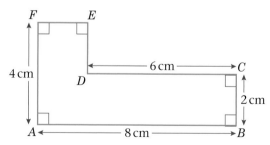

Find the distance of the centre of mass of the lamina from:

a AF,

b AB.

You can find the centre of mass in three different ways.

Method 1

You can summarise the area of each part of the shape and the positions of G_1 and G_2 in a table. Because the lamina is uniform you only need to know the area of each piece.

Area	8	12	20
x	1	5	$\bar{x}$
y	2	1	$\bar{y}$

The centre of mass of the first rectangle is (1, 2).

$$8\binom{1}{2} + 12\binom{5}{1} = 20\binom{\bar{x}}{\bar{y}}$$

The centre of mass of the second rectangle is at its centre (5, 1).

$$\binom{8}{16} + \binom{60}{12} = 20\binom{\bar{x}}{\bar{y}}$$

Using $\sum m_i \mathbf{r}_i = \bar{\mathbf{r}} \sum m_i$

$$\binom{68}{28} = 20\binom{\bar{x}}{\bar{y}}$$

Simplify.

$$3.4 = \bar{x}$$
$$1.4 = \bar{y}$$

Solve for $\bar{x}$ and $\bar{y}$.

Method 2

Split the shape using the dotted line shown.
The centre of the square is (1, 3).
The centre of the rectangle is (4, 1).

Area	4	16	20
x	1	4	$\bar{x}$
y	3	1	$\bar{y}$

$$4\binom{1}{3} + 16\binom{4}{1} = 20\binom{\bar{x}}{\bar{y}}$$

Using $\sum m_i \mathbf{r}_i = \bar{\mathbf{r}} \sum m_i$

$$\binom{4}{12} + \binom{64}{16} = 20\binom{\bar{x}}{\bar{y}}$$

$$\binom{68}{28} = 20\binom{\bar{x}}{\bar{y}}$$

$$3.4 = \bar{x}$$
$$1.4 = \bar{y}$$

As before.

Method 3

Area	32	12	20
x	4	5	$\bar{x}$
y	2	3	$\bar{y}$

You obtain the lamina by starting with a rectangle and removing another rectangle.
G_1 is (4, 2)
G_2 is (5, 3)

$$32\binom{4}{2} - 12\binom{5}{3} = 20\binom{\bar{x}}{\bar{y}}$$

Note the subtraction, since you are removing, not adding, the second rectangle.

$$\binom{128}{64} - \binom{60}{36} = 20\binom{\bar{x}}{\bar{y}}$$

$$\binom{68}{28} = 20\binom{\bar{x}}{\bar{y}}$$

Simplify.

$$3.4 = \bar{x}$$
$$1.4 = \bar{y}$$

As before.

a Distance from AF is 3.4 cm
b Distance from AB is 1.4 cm

Remember to give your answers in the form asked for.

Example 14

A uniform circular disc, centre O, of radius 5 cm has two circular holes cut in it, as shown in the diagram.

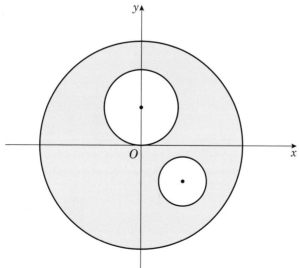

The larger hole has radius 2 cm and the smaller hole has radius 1 cm. The coordinates of the centres of the holes are (0, 2) and (2, −2) respectively. Find the coordinates of the centre of mass of the remaining lamina.

Area	$\pi \times 5^2$	$\pi \times 2^2$	$\pi \times 1^2$	$\pi(5^2 - 2^2 - 1^2)$
x	0	0	2	$\bar{x}$
y	0	2	-2	$\bar{y}$

You must use the 'removal' method. It is not possible to do this problem by treating the lamina as a sum of parts.

Setting out the key information in a table helps to clarify your working.

Note the subtraction signs for each removed section.

$$\pi 5^2 \begin{pmatrix} 0 \\ 0 \end{pmatrix} - \pi 2^2 \begin{pmatrix} 0 \\ 2 \end{pmatrix} - \pi 1^2 \begin{pmatrix} 2 \\ -2 \end{pmatrix} = \pi(5^2 - 2^2 - 1^2)\begin{pmatrix} \bar{x} \\ \bar{y} \end{pmatrix}$$

$$\begin{pmatrix} 0 \\ 0 \end{pmatrix} - \begin{pmatrix} 0 \\ 8 \end{pmatrix} - \begin{pmatrix} 2 \\ -2 \end{pmatrix} = 20\begin{pmatrix} \bar{x} \\ \bar{y} \end{pmatrix}$$

Cancel the π's and simplify.

$$\begin{pmatrix} -2 \\ -6 \end{pmatrix} = 20\begin{pmatrix} \bar{x} \\ \bar{y} \end{pmatrix}$$

Collect terms.

$$\begin{pmatrix} -0.1 \\ -0.3 \end{pmatrix} = \begin{pmatrix} \bar{x} \\ \bar{y} \end{pmatrix}$$

Solve for $\bar{x}$ and $\bar{y}$.

The coordinates of the centre of mass of the lamina are $(-0.1, -0.3)$.

Exercise 2D

The following diagrams show uniform plane figures. Each one is drawn on a grid of unit squares. Find, in each case, the coordinates of the centre of mass.

1
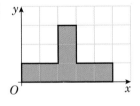

2
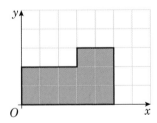

3
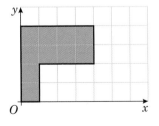

4
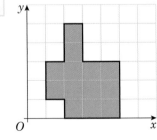

5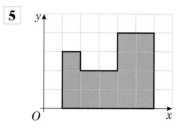

6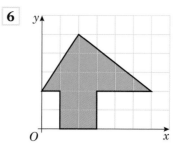

7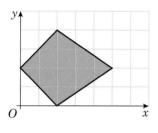

8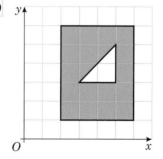

9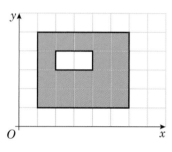

10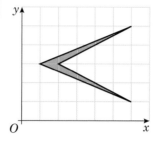

11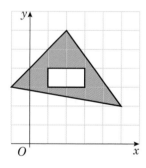

12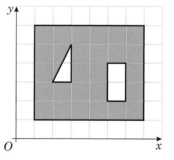

13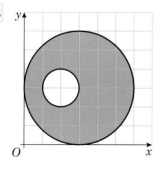

14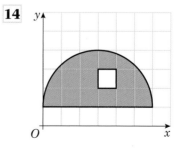

15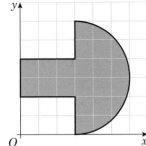

16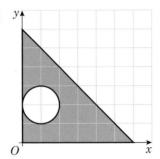

2.5 You can find the centre of mass of a framework by using the centre of mass of each rod or wire which makes up the framework.

Uniform circular arc

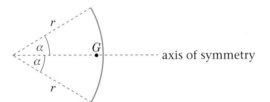

This result can be found in the formula booklet. You may need to use it in your M2 examination, although it will not be proved until M3.

A uniform circular arc of radius r and centre angle 2α, where α is measured in radians, has its centre of mass on the axis of symmetry at a distance $\frac{r \sin \alpha}{\alpha}$ from the centre.

■ **A framework** consists of a number of rods joined together or a number of pieces of wire joined together.

Provided that you can identify the position of the centre of mass of each of the rods or pieces of wire that make up a framework you can find the position of the centre of mass of the whole framework.

Example **15**

A framework consists of a uniform length of wire which has been bent into the shape of a letter L, as shown.

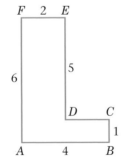

All lengths are in cm.

Find the distance of the centre of mass of the framework from AB and AF.

Since the wire is uniform, the mass of each edge will be proportional to its length. The centre of mass of each edge will be at its mid-point.

Taking A as the origin and axes along AB and AF:

$$4\begin{pmatrix} 2 \\ 0 \end{pmatrix} + 1\begin{pmatrix} 4 \\ 0.5 \end{pmatrix} + 2\begin{pmatrix} 3 \\ 1 \end{pmatrix} + 5\begin{pmatrix} 2 \\ 3.5 \end{pmatrix} + 2\begin{pmatrix} 1 \\ 6 \end{pmatrix} + 6\begin{pmatrix} 0 \\ 3 \end{pmatrix} = 20\begin{pmatrix} \bar{x} \\ \bar{y} \end{pmatrix}$$

$$\begin{pmatrix} 8 \\ 0 \end{pmatrix} + \begin{pmatrix} 4 \\ 0.5 \end{pmatrix} + \begin{pmatrix} 6 \\ 2 \end{pmatrix} + \begin{pmatrix} 10 \\ 17.5 \end{pmatrix} + \begin{pmatrix} 2 \\ 12 \end{pmatrix} + \begin{pmatrix} 0 \\ 18 \end{pmatrix} = 20\begin{pmatrix} \bar{x} \\ \bar{y} \end{pmatrix}$$

$$\begin{pmatrix} 30 \\ 50 \end{pmatrix} = 20\begin{pmatrix} \bar{x} \\ \bar{y} \end{pmatrix}$$

$$\begin{pmatrix} 1.5 \\ 2.5 \end{pmatrix} = \begin{pmatrix} \bar{x} \\ \bar{y} \end{pmatrix}$$

Distance from AB is 2.5 cm

Distance from AF is 1.5 cm

Each term on the LHS consists of the length of an edge multiplied by the position vector of the mid-point of the edge.

Simplify and collect terms.

This is $\bar{y}$.

This is $\bar{x}$.

Example 16

Find the position of the centre of mass of a framework constructed from a uniform piece of wire bent into the shape shown:

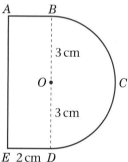

where the wire BCD is a semicircle, centre O, of radius 3 cm and wire $BAED$ forms three sides of a rectangle $ABDE$.

Take O as the origin and axes along OC and OB.

$$3\pi\begin{pmatrix} \frac{6}{\pi} \\ 0 \end{pmatrix} + 2\begin{pmatrix} -1 \\ -3 \end{pmatrix} + 2\begin{pmatrix} -1 \\ 3 \end{pmatrix} + 6\begin{pmatrix} -2 \\ 0 \end{pmatrix} = (10 + 3\pi)\begin{pmatrix} \bar{x} \\ \bar{y} \end{pmatrix}$$

$$\begin{pmatrix} 18 \\ 0 \end{pmatrix} + \begin{pmatrix} -2 \\ -6 \end{pmatrix} + \begin{pmatrix} -2 \\ 6 \end{pmatrix} + \begin{pmatrix} -12 \\ 0 \end{pmatrix} = (10 + 3\pi)\begin{pmatrix} \bar{x} \\ \bar{y} \end{pmatrix}$$

$$\begin{pmatrix} 2 \\ 0 \end{pmatrix} = (10 + 3\pi)\begin{pmatrix} \bar{x} \\ \bar{y} \end{pmatrix}$$

$$\bar{x} = \frac{2}{(10 + 3\pi)}$$

$$\bar{y} = 0$$

Use the result for the centre of mass of a uniform circular arc with $\alpha = \frac{\pi}{2}$ and $r = 3$ to find the centre of mass of the arc BCD. 3π is the length of the arc.

Simplify.

Collect terms.

You could have used the symmetry to deduce this without any working.

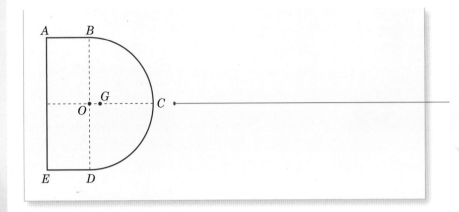

G is the centre of mass of the framework, on the axis of symmetry, a distance $\dfrac{2}{10 + 3\pi}$ cm to the right of O.

Exercise 2E

Find the coordinates of the centres of mass of the shapes shown in Exercise 2D questions **1** to **5** on page 48, regarding them as uniform plane wire frameworks.

6 Find the position of the centre of mass of the framework shown in the diagram which is formed by bending a uniform piece of wire of total length $(12 + 2\pi)$ cm to form a sector of a circle, centre O, radius 6 cm.

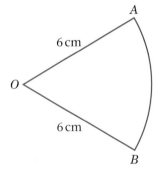

7 A uniform length of wire is bent to form the shape shown in the diagram:

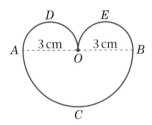

ACB is a semicircle of radius 3 cm, centre O.
ADO and BEO are both semicircles of radius 1.5 cm.
Find the position of the centre of mass of the framework.

2.6 You can solve problems involving a lamina in equilibrium.

A lamina can be suspended by means of a string attached to some point of the lamina, or can be allowed to pivot freely about horizontal axis which passes through some point of the lamina.

■ **When a lamina is suspended freely from a fixed point or pivots freely about a horizontal axis it will rest in equilibrium in a vertical plane with its centre of mass vertically below the point of suspension or the pivot.**

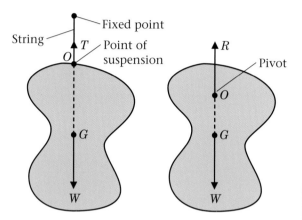

The first lamina is suspended from a fixed point. There are only two forces acting on it: the weight of the lamina and the tension in the string. Both forces pass through the point of suspension.

The second lamina is free to rotate about a fixed horizontal pivot. There are only two forces acting on it: the weight of the lamina and the reaction of the pivot on the lamina. Both pass through the point of suspension.

The resultant of the moments about O in both laminas is zero.

You also need to be able to answer questions about a lamina resting on an inclined plane.

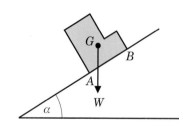

■ **If a lamina rests in equilibrium on a rough inclined plane then the line of action of the weight of the lamina must pass through the side of the lamina AB which is in contact with the plane.**

The weight of the lamina produces a clockwise moment about A which keeps the lamina in contact with the plane.

If the angle of the plane is increased so that the line of action of the weight passes outside the side AB then the weight produces an anticlockwise moment about A. The lamina will topple over.

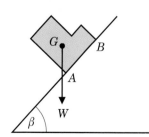

You can usually assume that the coefficient of friction between the lamina and the plane is large enough to prevent the lamina from slipping down the plane.

Example 17

Find the angle that the line AB makes with the vertical if this L-shaped uniform lamina is freely suspended from

a A,
b B,
c E.

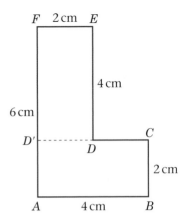

First find the centre of mass of the lamina.

Split the lamina along CD'.

Take A as the origin and axes along AB and AF,

$$8\binom{2}{1} + 8\binom{1}{4} = 16\binom{\bar{x}}{\bar{y}}$$

$$\binom{24}{40} = 16\binom{\bar{x}}{\bar{y}}$$

$$\bar{x} = 1.5$$

$$\bar{y} = 2.5$$

a

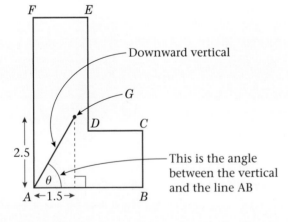

$$\tan \theta = \frac{2.5}{1.5}$$

$$\Rightarrow \theta = 59.0° \text{ (3 s.f.)}$$

Split the lamina into two rectangles.

Area $ABCD' = 8$
Area $DEFD' = 8$

Simplify.

You do not need to draw the lamina hanging.
Draw a line from the point of suspension to the centre of mass. Mark this in as the vertical.

θ is the angle required.

b

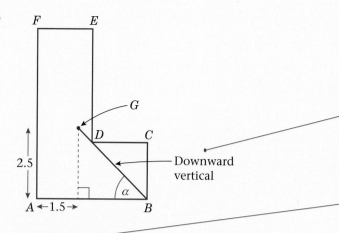

This time draw a line from B to G and mark this as the vertical.

α is the required angle.

Downward vertical

$$\tan \alpha = \frac{2.5}{4 - 1.5}$$

$$= \frac{2.5}{2.5}$$

$$\Rightarrow \theta = 45°$$

This time draw a line from E to G and mark this as the vertical.

c

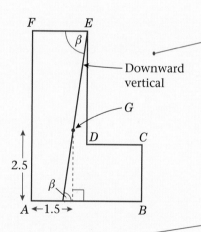

Downward vertical

β is the required angle.

$F\hat{E}G = \beta$ (alternate angles)

Using angle $F\hat{E}G$,

$$\tan \beta = \frac{AF - \bar{y}}{EF - \bar{x}}$$

$$\tan \beta = \frac{6 - 2.5}{2 - 1.5}$$

$$= \frac{3.5}{0.5}$$

$$\Rightarrow \beta = 81.9° \text{ (3 s.f.)}$$

Example 18

The L-shaped lamina from Example 17 is placed with *AB* in contact with a rough inclined plane. The angle of the plane is gradually increased. Assuming that the lamina does not slide down the plane, find the angle that the plane makes with the horizontal when the lamina is about to topple over.

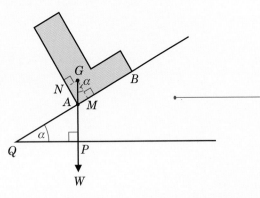

When the lamina is about to topple its centre of mass *G* will be vertically above the point *A.*

The plane must be rough enough to prevent the lamina from sliding down.

The moment of *W* about *A* is zero.

$AM = \bar{x} = 1.5; AN = \bar{y} = 2.5$

See Example 17.

Angle $G\hat{A}M = 90° - \alpha$

$G\hat{A}M = Q\hat{A}P$

so, angle $A\hat{G}M = \alpha$

In $\triangle AMG$, $\tan \alpha = \dfrac{AM}{MG} = \dfrac{\bar{x}}{\bar{y}}$

Since $\triangle AMG$ is right-angled.

$\tan \alpha = \dfrac{1.5}{2.5} = \dfrac{3}{5}$

Multiplying top and bottom by 2.

$\alpha = 31°$ (nearest degree)

> When a lamina on an inclined plane is about to topple its centre of mass will be vertically above the lowest point of the lamina which is contact with the plane.

Exercise 2F

1 a The lamina from question 1 in Exercise 2D is shown.

The lamina is freely suspended from the point *O* and hangs in equilibrium.
Find the angle between *OA* and the downward vertical.

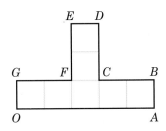

b The lamina from question 2 in Exercise 2D is shown below.

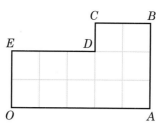

The lamina is freely suspended from the point O and hangs in equilibrium. Find the angle between OA and the downward vertical.

c The lamina from question 3 in Exercise 2D is shown below.

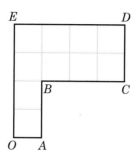

The lamina is freely suspended from the point O and hangs in equilibrium. Find the angle between OA and the downward vertical.

2 The lamina from question 4 in Exercise 2D is shown below.

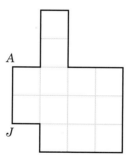

The lamina is freely suspended from the point A and hangs in equilibrium. Find the angle between AJ and the downward vertical.

3 The lamina from question 7 in Exercise 2D is shown below.

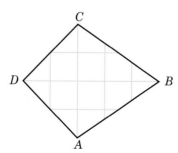

The lamina is free to rotate about a fixed smooth horizontal axis, perpendicular to the plane of the lamina, passing through the point A, and hangs in equilibrium. Find the angle between AC and the horizontal.

4 The framework in question 6, Exercise 2E is freely suspended from the point A and allowed to hang in equilibrium. Find the angle between OA and the downward vertical.

5 The shape in question 7, Exercise 2E is freely suspended from the point A and allowed to hang in equilibrium. Find the angle between OA and the horizontal.

6 The uniform triangular lamina ABC shown below is placed on a rough plane inclined at an angle α to the horizontal.

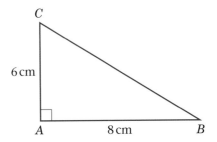

The edge AB is in contact with the plane, with A below B.
Given that the lamina is on the point of toppling about A, find the value of α.

7 $PQRS$ is a uniform lamina.

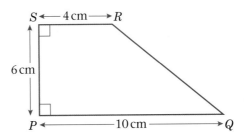

a Find the distance of the centre of mass of the lamina from
   i PS **ii** PQ.

b The diagram shows the lamina on a rough inclined plane of angle α.

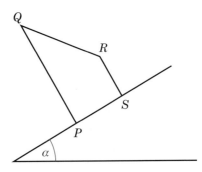

Given that the lamina is about to topple about the point P, find the value of α, giving your answer to 3 s.f.

Mixed exercise 2G

1 The diagram shows a uniform lamina consisting of a semi-circle joined to a triangle *ADC*.

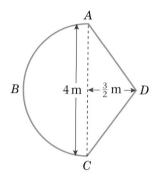

The sides *AD* and *DC* are equal.

a Find the distance of the centre of mass of the lamina from *AC*.

The lamina is freely suspended from *A* and hangs at rest.

b Find, to the nearest degree, the angle between *AC* and the vertical.

The mass of the lamina is *M*. A particle *P* of mass *kM* is attached to the lamina at *D*. When suspended from *A*, the lamina now hangs with its axis of symmetry, *BD*, horizontal.

c Find, to 3 s.f., the value of *k*.

2 A uniform triangular lamina *ABC* is in equilibrium, suspended from a fixed point *O* by a light inextensible string attached to the point *B* of the lamina, as shown in the diagram.

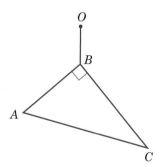

Given that $AB = 9$ cm, $BC = 12$ cm and $A\hat{B}C = 90°$, find the angle between *BC* and the downward vertical.

3 Four particles *P*, *Q*, *R* and *S* of masses 3 kg, 5 kg, 2 kg and 4 kg are placed at the points $(1, 6)$, $(-1, 5)$, $(2, -3)$ and $(-1, -4)$ respectively. Find the coordinates of the centre of mass of the particles.

4 A uniform rectangular piece of card *ABCD* has $AB = 3a$ and $BC = a$. One corner of the rectangle is folded over to form a trapezium *ABED* as shown in the diagram:

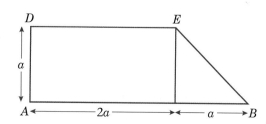

Find the distance of the centre of mass of the trapezium from

a *AD*,

b *AB*.

The lamina *ABED* is freely suspended from *E* and hangs at rest.

c Find the angle between *DE* and the horizontal.

The mass of the lamina is *M*. A particle of mass *m* is attached to the lamina at the point *B*. The lamina is freely suspended from *E* and it hangs at rest with *AB* horizontal.

d Find *m* in terms of *M*.

5 A thin uniform wire of length $5a$ is bent to form the shape *ABCD*, where $AB = 2a$, $BC = 2a$, $CD = a$ and *BC* is perpendicular to both *AB* and *CD*, as shown in the diagram:

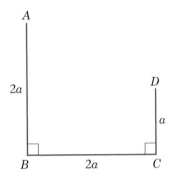

a Find the distance of the centre of mass of the wire from
  i *AB*,   **ii** *BC*.

The wire is freely suspended from *B* and hangs at rest.

b Find, to the nearest degree, the angle between *AB* and the vertical.

6 A uniform lamina consists of a rectangle *ABCD*, where $AB = 3a$ and $AD = 2a$, with a square hole *EFGA*, where $EF = a$, as shown in the diagram:

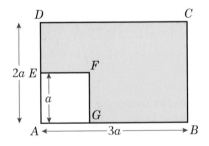

a Find the distance of the centre of mass of the lamina from
  i *AD*,   **ii** *AB*.

The lamina is balanced on a rough plane inclined to the horizontal at an angle θ. The plane of the lamina is vertical and the inclined plane is sufficiently rough to prevent the lamina from slipping. The side *GB* is in contact with the plane with *G* lower than *B*, as shown in the diagram:

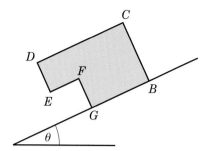

b Find, in degrees to 1 decimal place, the greatest value of θ for which the lamina can rest in equilibrium without toppling.

Summary of key points

1 The centre of mass of a large body is the point at which the whole mass of the body can be considered to be concentrated.

2 If a system of n particles with masses $m_1, m_2, ..., m_n$ are placed along the x-axis at the points $(x_1, 0), (x_2, 0), ... (x_n, 0)$ respectively, then

$$\sum m_i x_i = \bar{x} \sum m_i$$

where $(\bar{x}, 0)$ is the position of the centre of mass of the system.

3 If a system consists of n particles: mass m_1 with position vector $\mathbf{r}_1$, mass m_2 with position vector $\mathbf{r}_2$, ... mass m_n with position vector $\mathbf{r}_n$ then

$$\sum m_i \mathbf{r}_i = \bar{\mathbf{r}} \sum m_i$$

where $\bar{\mathbf{r}}$ is the position vector of the centre of mass of the system.

4 If a question does not specify axes or coordinates you will need to choose your own axes and origin.

5 An object which has one dimension (its thickness) very small compared with the other two (its length and width) is modelled as a lamina. This means that it is regarded as being two-dimensional with area rather than volume.

6 The centre of mass of a uniform triangular lamina is at the intersection of the medians. This point is called the centroid of the triangle.

7 If the coordinates of the three vertices of a uniform triangular lamina are $(x_1, y_1), (x_2, y_2)$ and (x_3, y_3) then the coordinates of the centre of mass are given by taking the average (mean) of the coordinates of the vertices:

G is the point $\left(\dfrac{x_1 + x_2 + x_3}{3}, \dfrac{y_1 + y_2 + y_3}{3} \right)$

8 The centre of mass of a uniform plane lamina will always lie on an axis of symmetry.

9 A framework consists of a number of rods joined together or a number of pieces of wire joined together.

10 When a lamina is suspended freely from a fixed point or pivots freely about a horizontal axis it will rest in equilibrium in a vertical plane with its centre of mass vertically below the point of suspension or the pivot.

11 If a lamina rests in equilibrium on a rough inclined plane then the line of action of the weight of the lamina must pass through the side of the lamina AB which is in contact with the plane.

After completing this chapter you should be able to

1 use the principle of conservation of mechanical energy and the work–energy principle to solve problems about a moving particle

2 solve problems about a moving vehicle including calculating the power developed by its engine.

Work, energy and power

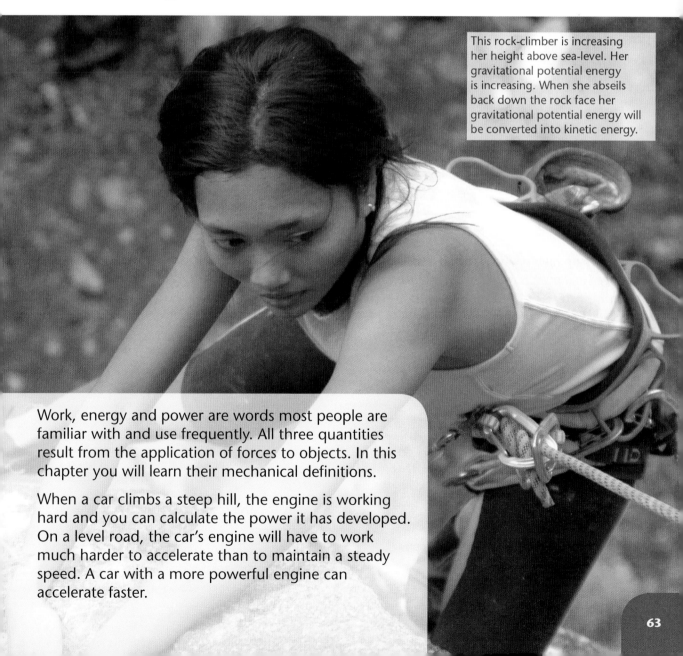

This rock-climber is increasing her height above sea-level. Her gravitational potential energy is increasing. When she abseils back down the rock face her gravitational potential energy will be converted into kinetic energy.

Work, energy and power are words most people are familiar with and use frequently. All three quantities result from the application of forces to objects. In this chapter you will learn their mechanical definitions.

When a car climbs a steep hill, the engine is working hard and you can calculate the power it has developed. On a level road, the car's engine will have to work much harder to accelerate than to maintain a steady speed. A car with a more powerful engine can accelerate faster.

3.1 **You can calculate the work done by a force when its point of application moves using the formula**

   work = force × distance moved by the point of application in the direction of the force.

When the force is measured in newtons and the distance in metres, the work done is measured in joules (J).

You can also calculate the work done against gravity when a particle is moved vertically. Work is done against gravity whenever a particle's vertical height is increased. This may be because the particle moved vertically or at an angle to the horizontal.

- **Work done against gravity = mgh, where m is the mass of the particle, g is the acceleration due to gravity and h is the vertical distance raised.**

This result is demonstrated in Example 3.

Example 1

A box is pulled 7 m across a horizontal floor by a horizontal force of magnitude 15 N. Calculate the work done by the force.

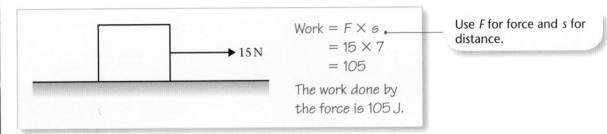

Work = $F \times s$

    = 15×7

    = 105

Use F for force and s for distance.

The work done by the force is 105 J.

Example 2

A packing case is pulled across a horizontal floor by a horizontal rope. The case moves at a constant speed and there is a constant resistance to motion of magnitude R newtons. When the case has moved a distance of 12 m the work done is 96 J. Calculate the magnitude of the resistance.

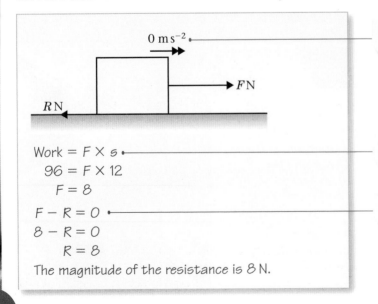

The case moves at a constant speed so its acceleration is $0\,\text{m s}^{-2}$.

Work = $F \times s$

  $96 = F \times 12$

    $F = 8$

$F - R = 0$

$8 - R = 0$

    $R = 8$

Use work = $F \times s$ to calculate the magnitude of the horizontal force.

Use $F = ma$ to calculate the value of R.

The magnitude of the resistance is 8 N.

Example 3

A bricklayer raises a load of bricks of total mass 30 kg at a constant speed by attaching a cable to the bricks. Assuming the cable is vertical, calculate the work done when the bricks are raised a distance of 7 m.

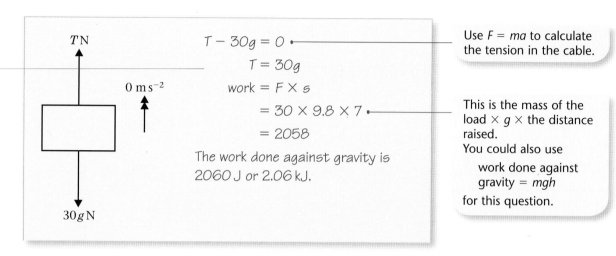

$T - 30g = 0$ ——— Use $F = ma$ to calculate the tension in the cable.

$T = 30g$

work $= F \times s$

$= 30 \times 9.8 \times 7$ ——— This is the mass of the load $\times g \times$ the distance raised.

$= 2058$

The work done against gravity is 2060 J or 2.06 kJ.

You could also use

work done against gravity $= mgh$

for this question.

Example 4

A package of mass 2 kg is pulled at a constant speed up a rough plane which is inclined at 30° to the horizontal. The coefficient of friction between the package and the surface is 0.35.
The package is pulled 12 m up a line of greatest slope of the plane. Calculate

a the work done against gravity,

b the work done against friction.

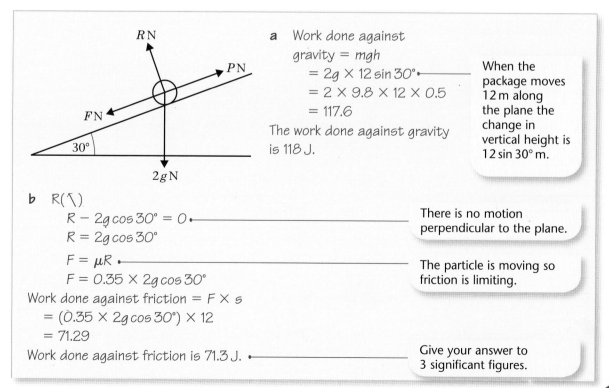

a Work done against gravity $= mgh$

$= 2g \times 12 \sin 30°$ ——— When the package moves 12 m along the plane the change in vertical height is $12 \sin 30°$ m.

$= 2 \times 9.8 \times 12 \times 0.5$

$= 117.6$

The work done against gravity is 118 J.

b $R(\nwarrow)$

$R - 2g \cos 30° = 0$ ——— There is no motion perpendicular to the plane.

$R = 2g \cos 30°$

$F = \mu R$ ——— The particle is moving so friction is limiting.

$F = 0.35 \times 2g \cos 30°$

Work done against friction $= F \times s$

$= (0.35 \times 2g \cos 30°) \times 12$

$= 71.29$

Work done against friction is 71.3 J. ——— Give your answer to 3 significant figures.

■ You can calculate the work done by a force acting at an angle to the direction of motion using the formula

$$\text{work done} = \frac{\text{component of force in}}{\text{direction of motion}} \times \frac{\text{distance moved}}{\text{in the same direction}}$$

Example 5

A sledge is pulled 15 m across a smooth sheet of ice by a force of magnitude 27 N. The force is inclined at 25° to the horizontal. By modelling the sledge as a particle calculate the work done by the force.

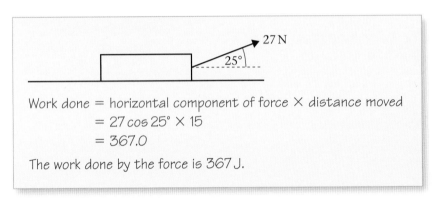

Work done = horizontal component of force × distance moved
= 27 cos 25° × 15
= 367.0

The work done by the force is 367 J.

Exercise 3A

Whenever a numerical value of g is required, take $g = 9.8\,\text{m s}^{-2}$.

1 Calculate the work done by a horizontal force of magnitude 0.6 N which pulls a particle a distance of 4.2 m across a horizontal floor.

2 A box is pulled 12 m across a smooth horizontal floor by a constant horizontal force. The work done by the force is 102 J. Calculate the magnitude of the force.

3 Calculate the work done against gravity when a particle of mass 0.35 kg is raised a vertical distance of 7 m.

4 A crate of mass 15 kg is raised through a vertical distance of 4 m. Calculate the work done against gravity.

5 A box is pushed 15 m across a horizontal surface. The box moves at a constant speed and the resistances to motion total 22 N. Calculate the work done by the force pushing the box.

6 A ball of mass 0.5 kg falls vertically 15 m from rest. Calculate the work done by gravity.

7 A cable is attached to a crate of mass 80 kg. The crate is raised vertically at a constant speed from the ground to the top of a building. The work done in raising the crate is 30 kJ. Calculate the height of the building.

8 A sledge is pulled 14 m across a horizontal sheet of ice by a rope inclined at 25° to the horizontal. The tension in the rope is 18 N and the ice can be assumed to be a smooth surface. Calculate the work done.

9 A parcel of mass 3 kg is pulled at a distance of 4 m across a rough horizontal floor. The parcel moves at a constant speed. The work done against friction is 30 J. Calculate the coefficient of friction between the parcel and the surface.

10 A block of wood of mass 2 kg is pushed across a rough horizontal floor. The block moves at $3\,m\,s^{-1}$ and the coefficient of friction between the block and the floor is 0.55. Calculate the work done in 2 seconds.

11 A girl of mass 52 kg climbs a vertical cliff which is 46 m high. Calculate the work she does against gravity.

12 A child of mass 25 kg slides 2 m down a smooth slope inclined at 35° to the horizontal. Calculate the work done by gravity.

13 A particle of mass 0.3 kg is pulled 2 m up a line of greatest slope of a plane which is inclined at 25° to the horizontal. Assuming that the particle moves along a line of greatest slope of the plane, calculate the work done against gravity.

14 A rough plane surface is inclined at an angle $\arcsin\frac{5}{13}$ to the horizontal. A packet of mass 8 kg is pulled at a constant speed up a line of greatest slope of the plane. The coefficient of friction between the packet and the plane is 0.3.

a Calculate the magnitude of the frictional force acting on the packet.

The packet moves a distance of 15 m up the plane. Calculate

b the work done against friction,

c the work done against gravity.

15 A particle P of mass 2 kg is projected up a line of greatest slope of a rough plane which is inclined at an angle $\arcsin\frac{3}{5}$ to the horizontal. The coefficient of friction between P and the plane is 0.35. The particle travels 3 m up the plane. Calculate the work done by friction.

16 A rough surface is inclined at an angle $\arcsin\frac{7}{25}$ to the horizontal. A particle of mass 0.5 kg is pulled 3 m at a constant speed up the surface by a force acting along a line of greatest slope. The only resistances to the motion are those due to friction and gravity. The work done by the force is 12 J. Calculate the coefficient of friction between the particle and the surface.

17 A rough surface is inclined at 40° to the horizontal. A particle of mass 1.5 kg is pulled at a constant speed up the surface by a force T acting along a line of greatest slope. The coefficient of friction between the particle and the surface is 0.4. Calculate the work done by T when the particle travels 8 m. You may assume that the only resistances to motion are due to gravity and friction.

3.2 You can calculate the kinetic energy of a moving particle and the potential energy of a particle.

■ Kinetic energy (K.E.) = $\frac{1}{2}mv^2$
where m is the mass of the particle and v is its velocity

■ Potential energy (P.E.) = mgh
where h is the height of the particle above an arbitrary fixed level

A particle possesses kinetic energy when it is moving. When the mass of the particle is measured in kilograms and its velocity is measured in metres per second the kinetic energy is measured in joules.

The work done by a force which accelerates a particle horizontally is connected to the kinetic energy of that particle.

■ Work done = change in kinetic energy

You can derive this result using formulae you already know:

$F = ma$ — Equation of motion

$v^2 = u^2 + 2as$ — Constant acceleration formula

$F = \dfrac{m(v^2 - u^2)}{2s}$ — Eliminate a from these two equations.

$Fs = \frac{1}{2}mv^2 - \frac{1}{2}mu^2$

Work done = final K.E. − initial K.E. — Fs is force × distance or work done.

Work done = increase in K.E.

Example 6

A particle of mass 0.3 kg is moving at a speed of $9\,\text{m s}^{-1}$. Calculate its kinetic energy.

K.E. = $\frac{1}{2}mv^2$

$= \frac{1}{2} \times 0.3 \times 9^2$

$= 12.15$

The K.E. of the particle is 12.2 J (3 s.f.).

Example 7

A box of mass 1.5 kg is pulled across a smooth horizontal surface by a horizontal force. The initial speed of the box is u m s^{-1} and its final speed is 3 m s^{-1}. The work done by the force is 1.8 J. Calculate the value of u.

> Work done $= \frac{1}{2}mv^2 - \frac{1}{2}mu^2$
>
> $1.8 = \frac{1}{2} \times 1.5 \times 3^2 - \frac{1}{2} \times 1.5u^2$
>
> $\frac{1}{2} \times 1.5u^2 = 4.95$
>
> $u^2 = \dfrac{4.95 \times 2}{1.5}$
>
> $u = 2.569$
>
> The initial speed of the box is 2.57 m s^{-1} (3 s.f.).

Example 8

A van of mass 2000 kg starts from rest at some traffic lights. After travelling 400 m the van's speed is 12 m s^{-1}. A constant resistance of 500 N acts on the van. Calculate the driving force which can be assumed to be constant.

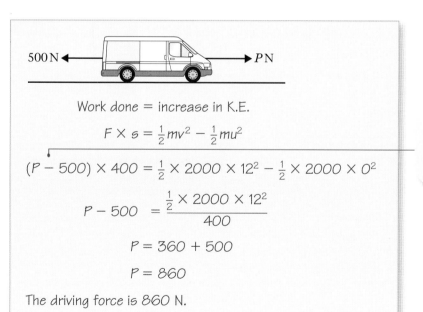

Work done $=$ increase in K.E.

$$F \times s = \frac{1}{2}mv^2 - \frac{1}{2}mu^2$$

$$(P - 500) \times 400 = \frac{1}{2} \times 2000 \times 12^2 - \frac{1}{2} \times 2000 \times 0^2$$

$$P - 500 = \dfrac{\frac{1}{2} \times 2000 \times 12^2}{400}$$

$$P = 360 + 500$$

$$P = 860$$

The driving force is 860 N.

> The force used to calculate the work done is the resultant force in the direction of the motion.

■ **You must choose a zero level before calculating a particle's potential energy.**

If the particle moves upwards its potential energy will increase. If it moves downwards its potential energy will decrease.

Example 9

A load of bricks of total mass 30 kg is lowered vertically to the ground through a distance of 15 m. Find the loss in potential energy.

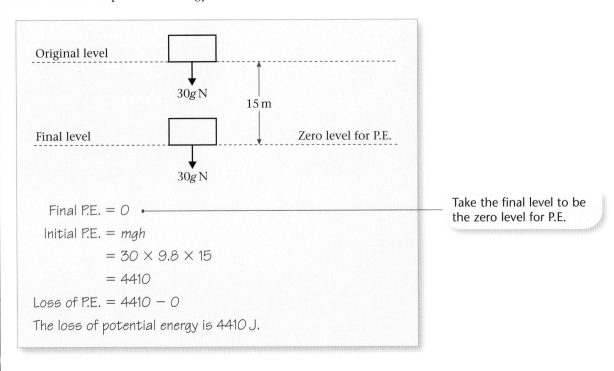

Take the final level to be the zero level for P.E.

Final P.E. = 0

Initial P.E. = mgh

$= 30 \times 9.8 \times 15$

$= 4410$

Loss of P.E. = $4410 - 0$

The loss of potential energy is 4410 J.

Example 10

A parcel of mass 3 kg is pulled 5 m up a plane inclined at an angle θ to the horizontal, where $\tan\theta = \frac{3}{4}$. Assuming that the parcel moves up a line of greatest slope of the plane, calculate the potential energy gained by the parcel.

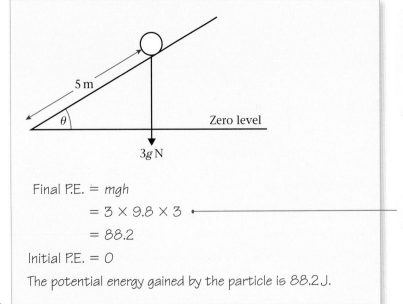

Draw a small triangle showing θ and use it to obtain the perpendicular height, h.

Final P.E. = mgh

$= 3 \times 9.8 \times 3$

$= 88.2$

The vertical distance moved by the particle is 3 m.

Initial P.E. = 0

The potential energy gained by the particle is 88.2 J.

Exercise 3B

Whenever a numerical value of g is required, take $g = 9.8 \, \text{m s}^{-2}$.

1 Calculate the kinetic energy of:
  a a particle of mass 0.3 kg moving at $15 \, \text{m s}^{-1}$
  b a particle of mass 3 kg moving at $2 \, \text{m s}^{-1}$
  c a box of mass 5 kg moving at $7.5 \, \text{m s}^{-1}$
  d an arrow of mass 0.5 kg moving at $200 \, \text{m s}^{-1}$
  e a boy of mass 25 kg running at $4 \, \text{m s}^{-1}$
  f a ball of mass 0.4 kg moving at $15 \, \text{m s}^{-1}$
  g a car of mass 800 kg moving at $20 \, \text{m s}^{-1}$

2 Find the change in potential energy of each of the following, stating in each case whether it is a loss or a gain:
  a a particle of mass 1.5 kg raised through a vertical distance of 3 m
  b a woman of mass 55 kg ascending a vertical distance of 15 m
  c a man of mass 75 kg descending a vertical distance of 30 m
  d a lift of mass 580 kg descending a vertical distance of 6 m
  e a man of mass 70 kg ascending a vertical distance of 36 m
  f a ball of mass 0.6 kg falling a vertical distance of 12 m
  g a lift of mass 800 kg ascending a vertical distance of 16 m

3 A particle of mass 1.2 kg decreases its speed from $12 \, \text{m s}^{-1}$ to $4 \, \text{m s}^{-1}$. Calculate the decrease in the particle's kinetic energy.

4 A van of mass 900 kg increases its speed from $5 \, \text{m s}^{-1}$ to $20 \, \text{m s}^{-1}$. Calculate the increase in the van's kinetic energy.

5 A particle of mass 0.2 kg increases its speed from $2 \, \text{m s}^{-1}$ to $v \, \text{m s}^{-1}$. The particle's kinetic energy increases by 6 J. Calculate the value of v.

6 An ice-skater of mass 45 kg is initially moving at $5 \, \text{m s}^{-1}$. She decreases her kinetic energy by 100 J. Calculate her final speed.

7 A playground slide is a plane inclined at 48° to the horizontal. A child of mass 25 kg slides down the slide for 4 m. Calculate the potential energy lost by the child.

8 A ball of mass 0.6 kg is dropped from a height of 2 m into a pond.
  a Calculate the kinetic energy of the ball as it hits the surface of the water.
  The ball begins to sink in the water with a speed of $4.8 \, \text{m s}^{-1}$.
  b Calculate the kinetic energy lost when the ball strikes the water.

9 A lorry of mass 2000 kg is initially travelling at 35 m s^{-1}. The brakes are applied, causing the lorry to decelerate at 1.2 m s^{-2} for 5 s. Calculate the loss of kinetic energy of the lorry.

10 A car of mass 750 kg moves along a stretch of road which can be modelled as a line of greatest slope of a plane inclined to the horizontal at 30°. As the car moves up the road for 500 m its speed reduces from 20 m s^{-1} to 15 m s^{-1}. Calculate

    a the loss of kinetic energy of the car,

    b the gain of potential energy of the car.

11 A man of mass 80 kg climbs a vertical cliff face of height h m. His potential energy increases by 15.7 kJ. Calculate the height of the cliff.

3.3 **You can use the principle of conservation of mechanical energy and the work–energy principle to solve problems involving a moving particle.**

■ **When no external forces (other than gravity) do work on a particle during its motion, the sum of the particle's kinetic and potential energies remains constant.**

> This is called the principle of conservation of mechanical energy.

This is true whether the particle moves vertically or along a path inclined to the horizontal.

The total energy possessed by a particle can only change during the particle's motion if some external force is doing work on the particle. Any non-gravitational resistance to motion acting on the particle will reduce the total energy of the particle as the particle will have to do work to overcome the resistance.

■ **The change in the total energy of a particle is equal to the work done on the particle.**

> This is called the work–energy principle.

Example 11

A smooth plane is inclined at 30° to the horizontal. A particle of mass 0.5 kg slides down a line of greatest slope of the plane. The particle starts from rest at point A and passes point B with a speed 6 m s^{-1}. Find the distance AB.

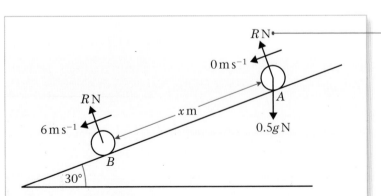

> Note that R does no work on the particle as it is always perpendicular to the motion.

Decrease in P.E. $= mgh$

$\qquad = 0.5 \times 9.8 \times (x \sin 30°)$

Increase in K.E. $= \frac{1}{2}mv^2 - \frac{1}{2}mu^2$

$\qquad = \frac{1}{2} \times 0.5 \times 6^2 - 0$

Decrease in P.E. = increase in K.E.

$0.5 \times 9.8 \times (x \sin 30°) = \frac{1}{2} \times 0.5 \times 6^2$

$$x = \frac{\frac{1}{2} \times 0.5 \times 6^2}{0.5 \times 9.8 \times \sin 30°}$$

$$x = 3.673$$

The distance AB is 3.67 m (3 s.f.).

The vertical distance moved by the particle is $x \sin 30°$, where x is the distance AB.

The only force acting on the particle which is doing work is gravity so you can apply the principle of conservation of mechanical energy.

Example 12

A particle of mass 2 kg is projected with speed $8 \, \text{m s}^{-1}$ up a line of greatest slope of a rough plane inclined at 45° to the horizontal. The coefficient of friction between the particle and the plane is 0.4. Calculate the distance the particle travels up the plane before coming to instantaneous rest.

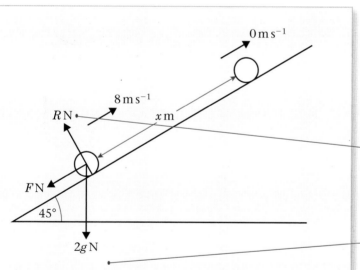

Note that R does no work on the particle as it is always perpendicular to the motion.

The particle has lost energy through having to work to overcome the frictional force.

Total loss of energy = K.E. lost − P.E. gained

$= \left(\frac{1}{2}mv^2 - \frac{1}{2}mu^2\right) - mgh$

$= \left(\frac{1}{2} \times 2 \times 8^2 - 0\right) - 2 \times 9.8 \times (x \sin 45°)$

$= 64 - 19.6x \sin 45°$

Work done against friction $= F \times x$

$R = 2g \cos 45°$

$F = \mu R = 0.4 \times 2g \cos 45° = 0.8g \cos 45°$

You need to find R so you can find F. Resolve perpendicular to the plane to find R.

The particle is moving so friction is limiting.

Loss of energy = work done against friction

$64 - 19.6x \sin 45° = 0.8g \cos 45° \times x$

$19.6x \sin 45° + 0.8gx \cos 45° = 64$

$x = \dfrac{64}{19.6 \sin 45° + 0.8g \cos 45°} = 3.298$

The particle moves 3.30 m (3 s.f.) up the plane.

> This is because of the work–energy principle.

Example **13**

A skier passes point A on a ski-run moving downhill at 6 m s^{-1}. After descending 50 m vertically the run begins to ascend. When the skier has ascended 25 m to point B her speed is 4 m s^{-1}. The skier and her skis have a combined mass of 55 kg. The total distance she travels from A to B is 1400 m. The non-gravitational resistances to motion are constant and have a total magnitude of 12 N. Calculate the work done by the skier.

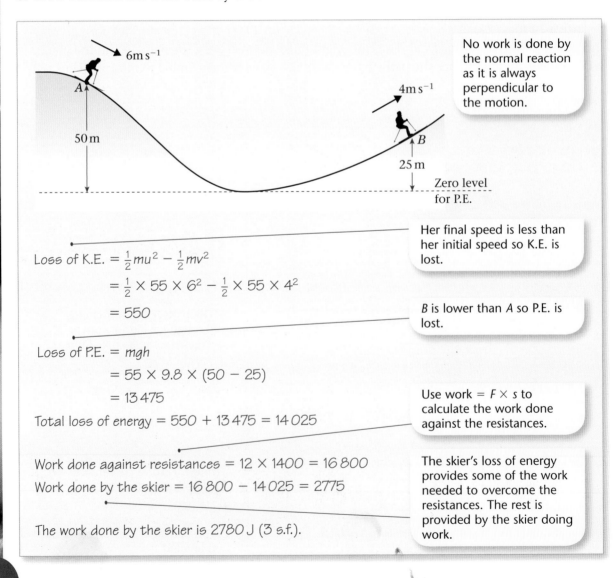

> No work is done by the normal reaction as it is always perpendicular to the motion.

Loss of K.E. $= \frac{1}{2}mu^2 - \frac{1}{2}mv^2$

$\qquad = \frac{1}{2} \times 55 \times 6^2 - \frac{1}{2} \times 55 \times 4^2$

$\qquad = 550$

> Her final speed is less than her initial speed so K.E. is lost.

Loss of P.E. $= mgh$

$\qquad = 55 \times 9.8 \times (50 - 25)$

$\qquad = 13\,475$

Total loss of energy $= 550 + 13\,475 = 14\,025$

> B is lower than A so P.E. is lost.

Work done against resistances $= 12 \times 1400 = 16\,800$

Work done by the skier $= 16\,800 - 14\,025 = 2775$

The work done by the skier is 2780 J (3 s.f.).

> Use work $= F \times s$ to calculate the work done against the resistances.

> The skier's loss of energy provides some of the work needed to overcome the resistances. The rest is provided by the skier doing work.

Exercise 3C

Whenever a numerical value of g is required, take $g = 9.8\,\text{m s}^{-2}$.

1 A particle of mass 0.4 kg falls a vertical distance of 7 m from rest.

  a Calculate the potential energy lost.

  b By assuming that air resistance can be neglected, calculate the final speed of the particle.

2 A stone of mass 0.5 kg is dropped from the top of a tower and falls vertically to the ground. It hits the ground with a speed of $12\,\text{m s}^{-1}$. Find

  a the kinetic energy gained by the stone,

  b the potential energy lost by the stone,

  c the height of the tower.

3 A box of mass 6 kg is pulled in a straight line across a smooth horizontal floor by a constant horizontal force of magnitude 10 N. The box has speed $2.5\,\text{m s}^{-1}$ when it passes through point P and speed $5\,\text{m s}^{-1}$ when it passes through point Q.

  a Find the increase in kinetic energy of the box.

  b Write down the work done by the force.

  c Find the distance PQ.

4 A particle of mass 0.4 kg moves in a straight line across a rough horizontal surface. The speed of the particle decreases from $8\,\text{m s}^{-1}$ to $4\,\text{m s}^{-1}$ as it travels 7 m.

  a Calculate the kinetic energy lost by the particle.

  b Write down the work done against friction.

  c Calculate the coefficient of friction between the particle and the surface.

5 A box of mass 3 kg is projected from point A of a rough horizontal floor with speed $6\,\text{m s}^{-1}$. The box moves in a straight line across the floor and comes to rest at point B. The coefficient of friction between the box and the floor is 0.4.

  a Calculate the kinetic energy lost by the box.

  b Write down the work done against friction.

  c Calculate the distance AB.

6 A particle of mass 0.8 kg falls a vertical distance of 5 m from rest. By considering energy, find the speed of the particle as it hits the ground. (You may assume that air resistance can be neglected.)

7 A stone of mass 0.3 kg is dropped from the top of a vertical cliff and falls freely under gravity. It hits the ground below with a speed of $20\,\text{m s}^{-1}$. Use energy considerations to calculate the height of the cliff.

8 A particle of mass 0.3 kg is projected vertically upwards and moves freely under gravity. The initial speed of the particle is $u\,\text{m s}^{-1}$. When the particle is 5 m above the point of projection its kinetic energy is 2.1 J. Calculate the value of u.

9 A bullet of mass 0.1 kg travelling at 500 m s^{-1} horizontally hits a vertical wall. The bullet penetrates the wall to a depth of 50 mm. The resistive force exerted on the bullet by the wall is constant. Calculate the magnitude of the resistive force.

10 A bullet of mass 150 g travelling at 500 m s^{-1} horizontally hits a vertical wall. The wall exerts a constant resistance of magnitude 250 000 N on the bullet. Calculate the distance the bullet penetrates the wall.

11 A package of mass 5 kg is released from rest and slides 2 m down a line of greatest slope of a smooth plane inclined at 35° to the horizontal.

   a Calculate the potential energy lost by the package.

   b Write down the kinetic energy gained by the package.

   c Calculate the final speed of the package.

12 A particle of mass 0.5 kg is released from rest and slides down a line of greatest slope of a smooth plane inclined at 30° to the horizontal. When the particle has moved a distance x m, its speed is 2 m s^{-1}. Find the value of x.

13 A particle of mass 0.2 kg is projected with speed 9 m s^{-1} up a line of greatest slope of a smooth plane inclined at 30° to the horizontal. The particle travels a distance x m before first coming to rest. By considering energy, calculate the value of x.

14 A particle of mass 0.6 kg is projected up a line of greatest slope of a smooth plane inclined at 40° to the horizontal. The particle travels 5 m before first coming to rest. Use energy considerations to calculate the speed of projection.

15 A box of mass 2 kg is projected with speed 6 m s^{-1} up a line of greatest slope of a rough plane inclined at 30° to the horizontal. The coefficient of friction between the box and the plane is $\frac{1}{3}$. Use the work–energy principle to calculate the distance the box travels up the plane before first coming to rest.

16 A cyclist freewheels down a hill inclined at 30° to the horizontal. The cyclist and his cycle have a combined mass of 80 kg. His speed increases from 3 m s^{-1} to 12 m s^{-1}. Assuming that resistances can be ignored, calculate

   a the potential energy lost by the cyclist,

   b the distance travelled by the cyclist.

17 A cyclist starts from rest and freewheels down a hill inclined at 20° to the horizontal. After travelling 60 m the road becomes horizontal and the cyclist travels a further 50 m before coming to rest. The cyclist and her cycle have a combined mass of 70 kg and the resistance to motion remains constant throughout. Calculate the magnitude of the resistance.

18 A girl and her sledge have a combined mass of 40 kg. She starts from rest and descends a slope which is inclined at 25° to the horizontal. At the bottom of the slope the ground becomes horizontal for 15 m before rising at 6° to the horizontal. The girl travels 25 m up the slope before coming to rest once more. There is a constant resistance to motion of magnitude 18 N. Calculate the distance the girl travels down the slope.

3.4 You can calculate the power developed by an engine and solve problems about moving vehicles.

■ **Power is the rate of doing work.**

Power is measured in watts (W) where **1 watt is 1 joule per second**. The power of an engine is often given in kilowatts (kW).

The power developed by the engine of a moving vehicle is calculated using the following formula

■ **Power = $F \times v$ where F is the driving force produced by the engine and v is the speed of the vehicle**
When the force is in newtons and the speed is in metres per second, the power is in watts.

Example 14

A truck is being pulled up a slope at a constant speed of 8 m s^{-1} by a force of magnitude 2000 N acting parallel to the direction of motion of the truck. Calculate, in kilowatts, the power developed.

Work done per second = (2000×8) = 16 000 J   Work done per second = force $\times$ distance moved per second.

Power = rate of doing work = 16 000 W

The power developed is 16 kW   1 kW = 1000 W.

Example 15

A van of mass 1250 kg is travelling along a horizontal road. The van's engine is working at 24 kW. The constant resistance to motion has magnitude 600 N. Calculate

a the acceleration of the van when it is travelling at $6 \, \text{m s}^{-1}$,

b the maximum speed of the van.

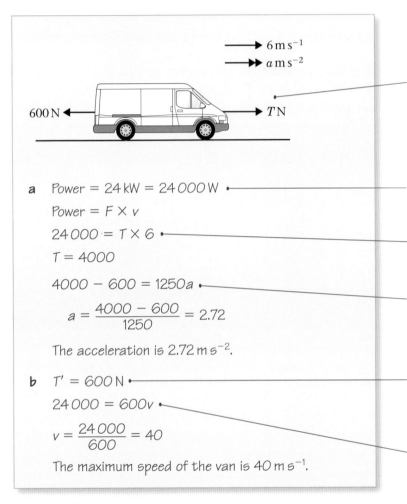

We often use T for the tractive (or pulling) force.

a Power = 24 kW = 24 000 W

You must work with power in watts.

Power = $F \times v$

$24\,000 = T \times 6$

Use power = $F \times v$ to find the driving force.

$T = 4000$

$4000 - 600 = 1250a$

Use $F = ma$ to find the acceleration.

$$a = \frac{4000 - 600}{1250} = 2.72$$

The acceleration is $2.72 \, \text{m s}^{-2}$.

b $T' = 600 \, \text{N}$

At maximum speed there will be no acceleration, so the resultant horizontal force will be zero.

$24\,000 = 600v$

$$v = \frac{24\,000}{600} = 40$$

Use power = $F \times v$ to find the speed.

The maximum speed of the van is $40 \, \text{m s}^{-1}$.

Example 16

A car of mass 1100 kg is travelling at a constant speed of 15 m s^{-1} along a straight road which is inclined at 7° to the horizontal. The engine is working at a rate of 24 kW.

a Calculate the magnitude of the non-gravitational resistances to motion.

The rate of working of the engine is now increased to 28 kW.

Assuming the resistances to motion are unchanged,

b calculate the initial acceleration of the car.

a

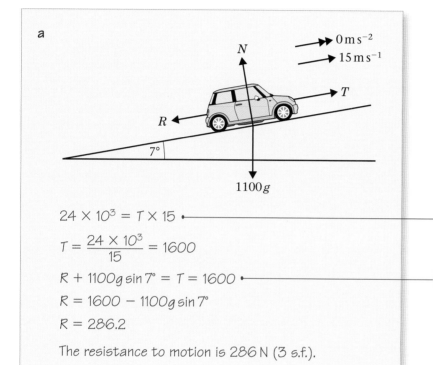

$$24 \times 10^3 = T \times 15$$ ●————— Use power = $F \times v$ to find T.

$$T = \frac{24 \times 10^3}{15} = 1600$$

$$R + 1100g \sin 7° = T = 1600$$ ●————— Resolve along the slope to find the resistance.

$$R = 1600 - 1100g \sin 7°$$

$$R = 286.2$$

The resistance to motion is 286 N (3 s.f.).

b

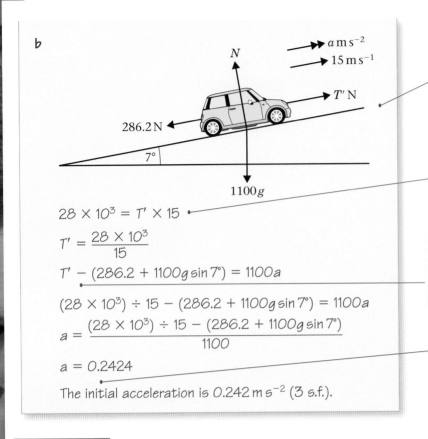

$286.2\,\text{N}$

N

$a\,\text{m}\,\text{s}^{-2}$

$15\,\text{m}\,\text{s}^{-1}$

$T'\,\text{N}$

$7°$

$1100g$

Draw a new diagram for the new situation. As the power has changed, the driving force and the acceleration will change.

Use power = $F \times v$ to find the new driving force. Initially, the speed will remain at $15\,\text{m}\,\text{s}^{-1}$.

$28 \times 10^3 = T' \times 15$

$T' = \dfrac{28 \times 10^3}{15}$

$T' - (286.2 + 1100g \sin 7°) = 1100a$

$(28 \times 10^3) \div 15 - (286.2 + 1100g \sin 7°) = 1100a$

$a = \dfrac{(28 \times 10^3) \div 15 - (286.2 + 1100g \sin 7°)}{1100}$

$a = 0.2424$

The initial acceleration is $0.242\,\text{m}\,\text{s}^{-2}$ (3 s.f.).

Use $F = ma$ to find the acceleration. Remember to use at least 4 significant figures for previously calculated answers.

We need 'initial' here as once the car accelerates its speed increases so the power and/or the driving force must also change.

Exercise 3D

Whenever a numerical value of g is required, take $g = 9.8\,\text{m}\,\text{s}^{-2}$.

1 A force of $1500\,\text{N}$ pulls a van up a slope at a constant speed of $12\,\text{m}\,\text{s}^{-1}$. Calculate, in kW, the power developed.

2 A car is travelling at $15\,\text{m}\,\text{s}^{-1}$ and its engine is producing a driving force of $1000\,\text{N}$. Calculate the power developed.

3 The engine of a van is working at $5\,\text{kW}$ and the van is travelling at $18\,\text{m}\,\text{s}^{-1}$. Find the magnitude of the driving force produced by the van's engine.

4 A car's engine is working at $15\,\text{kW}$. The car is travelling along a horizontal road. The total resistance to motion has a magnitude of $600\,\text{N}$. Calculate the maximum speed of the car.

5 A car has a maximum speed of $40\,\text{m}\,\text{s}^{-1}$ when travelling along a horizontal road against a constant resistance of $500\,\text{N}$. Calculate the power the car's engine must develop to maintain this speed.

6 A van is travelling along a horizontal road at a constant speed of $16\,\text{m}\,\text{s}^{-1}$. The van's engine is working at $8.8\,\text{kW}$. Calculate the magnitude of the resistance to motion.

7 A car of mass $850\,\text{kg}$ is travelling along a straight horizontal road against resistances totalling $350\,\text{N}$. The car's engine is working at $9\,\text{kW}$. Calculate

   a the acceleration when the car is travelling at $7\,\text{m}\,\text{s}^{-1}$,

   b the acceleration when the car is travelling at $15\,\text{m}\,\text{s}^{-1}$,

   c the maximum speed of the car.

8 A car of mass 900 kg is travelling along a straight horizontal road at a speed of $20\,\text{m}\,\text{s}^{-1}$. The constant resistances to motion total 300 N. The car is accelerating at $0.3\,\text{m}\,\text{s}^{-2}$. Calculate the power developed by the engine.

9 A car of mass 1000 kg is travelling along a straight horizontal road. The car's engine is working at 12 kW. When its speed is $24\,\text{m}\,\text{s}^{-1}$ its acceleration is $0.2\,\text{m}\,\text{s}^{-2}$. The resistances to motion have a total magnitude of R newtons. Calculate the value of R.

10 A cyclist is travelling along a straight horizontal road. The resistance to his motion is constant and has magnitude 28 N. The maximum rate at which he can work is 280 W. Calculate his maximum speed.

11 A van of mass 1200 kg is travelling up a straight road inclined at 5° to the horizontal. The van moves at a constant speed of $20\,\text{m}\,\text{s}^{-1}$ and its engine is working at 24 kW. The resistance to motion from non-gravitational forces has magnitude R newtons.

  a Calculate the value of R.

The road now becomes horizontal. The resistance to motion from non-gravitational forces is unchanged.

  b Calculate the initial acceleration of the car.

12 A car of mass 800 kg is travelling at $18\,\text{m}\,\text{s}^{-1}$ along a straight horizontal road. The car's engine is working at a constant rate of 26 kW against a constant resistance of magnitude 750 N.

  a Find the acceleration of the car.

The car now ascends a straight road, inclined at 9° to the horizontal. The resistance to motion from non-gravitational forces is unchanged and the car's engine works at the same rate.

  b Find the maximum speed at which the car can travel up the road.

13 A van of mass 1500 kg is travelling at its maximum speed of $30\,\text{m}\,\text{s}^{-1}$ along a straight horizontal road against a constant resistance of magnitude 600 N.

  a Find the power developed by the van's engine.

The van now travels up a straight road inclined at 8° to the horizontal. The van's engine works at the same rate and the resistance to motion from non-gravitational forces is unchanged.

  b Find the maximum speed at which the van can ascend the road.

14 A car is moving along a straight horizontal road with speed $v\,\text{m}\,\text{s}^{-1}$. The magnitude of the resistance to motion of the car is given by the formula $(150 + 3v)\,\text{N}$. The car's engine is working at 10 kW. Calculate the maximum value of v.

15 A train of mass 150 tonnes is moving up a straight track which is inclined at 2° to the horizontal. The resistance to the motion of the train from non-gravitational forces has magnitude 6 kN and the train's engine is working at a constant rate of 350 kW.

  a Calculate the maximum speed of the train.

The track now becomes horizontal. The engine continues to work at 350 kW and the resistance to motion remains 6 kN.

  b Find the initial acceleration of the train.

Mixed exercise 3E

Whenever a numerical value of g is required, take $g = 9.8 \, \text{m s}^{-2}$.

1 A cyclist and her bicycle have a combined mass of 70 kg. She is cycling at a constant speed of $6 \, \text{m s}^{-1}$ on a straight road up a hill inclined at 5° to the horizontal. She is working at a constant rate of 480 W. Calculate the magnitude of the resistance to motion from non-gravitational forces.

2 A boy hauls a bucket of water through a vertical distance of 25 m. The combined mass of the bucket and water is 12 kg. The bucket starts from rest and finishes at rest.

  a Calculate the work done by the boy.

  The boy takes 30 s to raise the bucket.

  b Calculate the average rate of working of the boy.

3 A particle P of mass 0.5 kg is moving in a straight line from A to B on a rough horizontal plane. At A the speed of P is $12 \, \text{m s}^{-1}$, and at B its speed is $8 \, \text{m s}^{-1}$. The distance from A to B is 25 m. The only resistance to motion is the friction between the particle and the plane. Find

  a the work done by friction as P moves from A to B,

  b the coefficient of friction between the particle and the plane.

4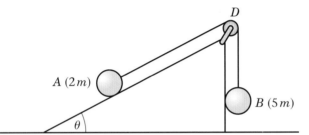

The diagram shows a particle A of mass $2m$ which can move on the rough surface of a plane inclined at an angle θ to the horizontal, where $\sin \theta = \frac{3}{5}$. A second particle B of mass $5m$ hangs freely attached to a light inextensible string which passes over a smooth light pulley fixed at D. The other end of the string is attached to A. The coefficient of friction between A and the plane is $\frac{3}{8}$. Particle B is initially hanging 2 m above the ground and A is 4 m from D. When the system is released from rest with the string taut A moves up a line of greatest slope of the plane.

  a Find the initial acceleration of A.

  When B has descended 1 m the string breaks.

  b By using the principle of conservation of energy calculate the total distance moved by A before it first comes to rest.

5 A car of mass 800 kg is travelling along a straight horizontal road. The resistance to motion from non-gravitational forces has a constant magnitude of 500 N. The engine of the car is working at a rate of 16 kW.

  a Calculate the acceleration of the car when its speed is $15 \, \text{m s}^{-1}$.

  The car comes to a hill at the moment when it is travelling at $15 \, \text{m s}^{-1}$. The road is still straight but is now inclined at 5° to the horizontal. The resistance to motion from non-gravitational forces is unchanged. The rate of working of the engine is increased to 24 kW.

  b Calculate the new acceleration of the car.

6 A car of mass 750 kg is moving at a constant speed of 18 m s^{-1} down a straight road inclined at an angle θ to the horizontal, where $\tan \theta = \frac{1}{20}$. The resistance to motion from non-gravitational forces has a constant magnitude of 1000 N.

a Find, in kW, the rate of working of the car's engine.

The engine of the car is now switched off and the car comes to rest T seconds later. The resistance to motion from non-gravitational forces is unchanged.

b Find the value of T.

7

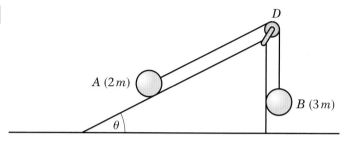

The diagram shows a particle A of mass $2m$ which can move on the rough surface of a plane inclined at an angle θ to the horizontal, where $\sin \theta = \frac{3}{5}$. A second particle B of mass $3m$ hangs freely attached to a light inextensible string which passes over a smooth pulley fixed at D. The other end of the string is attached to A. The coefficient of friction between A and the plane is $\frac{1}{4}$. The system is released from rest with the string taut and A moves up a line of greatest slope of the plane. When each particle has moved a distance s, A has not reached the pulley and B has not reached the ground.

a Find an expression for the potential energy lost by the system when each particle has moved a distance s.

When each particle has moved a distance s they are moving with speed v.

b Find an expression for v^2, in terms of s.

8 A parcel of mass 5 kg is resting on a platform inclined at 25° to the horizontal. The coefficient of friction between the parcel and the platform is 0.3. The parcel is released from rest and slides down a line of greatest slope of the platform. Calculate

a the speed of the parcel after it has been moving for 2 s,

b the potential energy lost by the parcel during this time.

9 A lorry of mass 16 000 kg is travelling up a straight road inclined at 12° to the horizontal. The lorry is travelling at a constant speed of 14 m s^{-1} and the resistance to motion from non-gravitational forces has a constant magnitude of 200 kN. Find the work done in 10 s by the engine of the lorry.

10 A particle P of mass 0.3 kg is moving in a straight line on a smooth horizontal surface under the action of a constant horizontal force. The particle passes point A with speed 6 m s^{-1} and point B with speed 12 m s^{-1}.

a Find the kinetic energy gained by P while moving from A to B.

b Write down the work done by the constant force.

The distance from A to B is 4 m.

c Calculate the magnitude of the force.

11 A box of mass 5 kg slides in a straight line across a rough horizontal floor. The initial speed of the box is $10\,\mathrm{m\,s^{-1}}$. The only resistance to the motion is the frictional force between the box and the floor. The box comes to rest after moving 8 m. Calculate

  a the kinetic energy lost by the box in coming to rest,

  b the coefficient of friction between the box and the floor.

12 A car of mass 900 kg is moving along a straight horizontal road. The resistance to motion has a constant magnitude. The engine of the car is working at a rate of 15 kW. When the car is moving with speed $20\,\mathrm{m\,s^{-1}}$, the acceleration of the car is $0.3\,\mathrm{m\,s^{-2}}$.

  a Find the magnitude of the resistance.

The car now moves downhill on a straight road inclined at 4° to the horizontal. The engine of the car is now working at a rate of 8 kW. The resistance to motion from non-gravitational forces remains unchanged.

  b Calculate the speed of the car when its acceleration is $0.5\,\mathrm{m\,s^{-2}}$.

13 A block of wood of mass 4 kg is pulled across a rough horizontal floor by a rope inclined at 15° to the horizontal. The tension in the rope is constant and has magnitude 75 N. The coefficient of friction between the block and the floor is $\frac{3}{8}$.

  a Find the magnitude of the frictional force opposing the motion.

  b Find the work done by the tension when the block moves 6 m.

The block is initially at rest.

  c Find the speed of the block when it has moved 6 m.

14 The engine of a lorry works at a constant rate of 20 kW. The lorry has a mass of 1800 kg. When moving along a straight horizontal road there is a constant resistance to motion of magnitude 600 N. Calculate

  a the maximum speed of the lorry,

  b the acceleration of the lorry, in $\mathrm{m\,s^{-2}}$, when its speed is $20\,\mathrm{m\,s^{-1}}$.

15 A car of mass 1200 kg is travelling at a constant speed of $20\,\mathrm{m\,s^{-1}}$ along a straight horizontal road. The constant resistance to motion has magnitude 600 N.

  a Calculate the power, in kW, developed by the engine of the car.

The rate of working of the engine of the car is suddenly increased and the initial acceleration of the car is $0.5\,\mathrm{m\,s^{-2}}$. The resistance to motion is unchanged.

  b Find the new rate of working of the engine of the car.

The car now comes to a hill. The road is still straight but is now inclined at 20° to the horizontal. The rate of working of the engine of the car is increased further to 50 kW. The resistance to motion from non-gravitational forces still has magnitude 600 N. The car climbs the hill at a constant speed $V\,\mathrm{m\,s^{-1}}$.

  c Find the value of V.

Summary of key points

1   For a force acting in the direction of the motion:

work = force × distance moved in the direction of the force

When the force is measured in newtons and the distance in metres, the work done is measured in joules (J).

2   Work done against gravity = mgh, where m is the mass of the particle, g is the acceleration due to gravity and h is the vertical distance raised.

3   You can calculate the work done by a force acting at an angle to the direction of motion using the formula:

$$\text{work done} = \frac{\text{component of force in}}{\text{direction of motion}} \times \frac{\text{distance moved in the}}{\text{same direction}}$$

4   The kinetic energy (K.E.) of a particle of mass m moving with speed $v\,\mathrm{m\,s^{-1}}$ is calculated using the formula

K.E. $= \frac{1}{2}mv^2$

For a mass in kg and a velocity in $\mathrm{m\,s^{-1}}$ the K.E. is measured in joules (J). K.E. is never negative.

5   The potential energy (P.E.) of a particle of mass m at a height h above a chosen fixed level is calculated using the formula

P.E. $= mgh$

Potential energy is also measured in joules. P.E. can be negative. You must choose a zero level before calculating a particle's potential energy.

6   **Principle of conservation of mechanical energy**
When no external forces (other than gravity) do work on a particle during its motion, the sum of the particle's kinetic and potential energies remains constant.

7   **Work–energy principle**
The change in the total energy of a particle is equal to the work done on the particle.

8   Power is the rate of doing work. For a moving particle:

power = $F \times v$, where F is the driving force in N and v is the speed in $\mathrm{m\,s^{-1}}$.

Power is measured in watts, where 1 watt (W) is 1 joule per second, or kilowatts, where 1 kilowatt (kW) is 1000 watts.

Review Exercise

Whenever a numerical value of g is required, take $g = 9.8\,\text{m s}^{-2}$.

1 A stone was thrown with velocity $20\,\text{m s}^{-1}$ at an angle of elevation of 30° from the top of a vertical cliff. The stone moved freely under gravity and reached the sea 5 s after it was thrown. Find

  a the vertical height above the sea from which the stone was thrown,

  b the horizontal distance covered by the stone from the instant when it was thrown until it reached the sea,

  c the magnitude and direction of the velocity of the stone when it reached the sea. **E**

2 A darts player throws darts at a dart board which hangs vertically. The motion of a dart is modelled as that of a particle moving freely under gravity. The darts move in a vertical plane which is perpendicular to the plane of the dart board. A dart is thrown horizontally with speed $12.6\,\text{m s}^{-1}$. It hits the board at a point which is 10 cm below the level from which it was thrown.

  a Find the horizontal distance from the point where the dart was thrown to the dart board.

The darts player moves his position. He now throws a dart from a point which is at a horizontal distance of 2.5 m from the dart board. He throws the dart at an angle of elevation α to the horizontal where $\tan \alpha = \frac{7}{24}$. The dart hits the board at a point which is at the same level as the point from which it was thrown.

  b Find the speed with which the dart was thrown. **E**

3 A particle is projected with velocity $(8\mathbf{i} + 10\mathbf{j})\,\text{m s}^{-1}$, where $\mathbf{i}$ and $\mathbf{j}$ are unit vectors horizontally and vertically respectively, from a point O at the top of a cliff and moves freely under gravity. Six seconds after projection, the particle strikes the sea at the point S. Calculate

  a the horizontal distance between O and S,

  b the vertical distance between O and S.

At time T seconds after projection, the particle is moving with velocity $(8\mathbf{i} - 14.5\mathbf{j})\,\mathrm{m\,s}^{-1}$.

c Find the value of T and the position vector, relative to O, of the particle at this instant. **(E)**

(4)

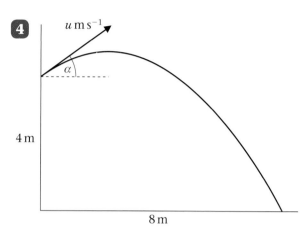

4 m

8 m

A ball is thrown from a point 4 m above horizontal ground. The ball is projected at an angle α above the horizontal, where $\tan \alpha = \frac{3}{4}$. The ball hits the ground at a point which is a horizontal distance 8 m from its point of projection, as shown in the figure above. The initial speed of the ball is $u\,\mathrm{m\,s}^{-1}$ and the time of flight is T seconds.

a Prove that $uT = 10$.

b Find the value of u.

As the ball hits the ground, its direction of motion makes an angle ϕ with the horizontal.

c Find $\tan \phi$. **(E)**

(5) A vertical cliff is 73.5 m high. Two stones A and B are projected simultaneously. Stone A is projected horizontally from the top of a cliff with speed $28\,\mathrm{m\,s}^{-1}$. Stone B is projected from the bottom of the cliff with speed $35\,\mathrm{m\,s}^{-1}$ at an angle α above the horizontal. The stones move freely under gravity in the same vertical plane and collide in mid-air.

a By considering the horizontal motion of each stone, prove that $\cos \alpha = \frac{4}{5}$.

b Find the time which elapses between the instant when the stones are projected and the instant when they collide. **(E)**

(6)

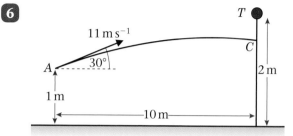

The object of a game is to throw a ball B from a point A to hit a target T which is placed at the top of a vertical pole, as shown in the figure above. The point A is 1 m above horizontal ground and the height of the pole is 2 m. The pole is a horizontal distance of 10 m from A. The ball B is projected from A with speed $11\,\mathrm{m\,s}^{-1}$ at an angle of elevation of 30°. The ball hits the pole at C. The ball B and the target T are modelled as particles.

a Calculate, to 2 decimal places, the time taken for B to move from A to C.

b Show that C is approximately 0.63 m below T.

The ball is thrown again from A. The speed of projection of B is increased to $V\,\mathrm{m\,s}^{-1}$, the angle of elevation remaining 30°. This time B hits T.

c Calculate the value of V.

d Explain why, in practice, a range of values of V would result in B hitting the target. **(E)**

(7) A particle P, projected from a point O on horizontal ground, moves freely under gravity and hits the ground again at A. Taking O as origin, OA as the x-axis and the upward vertical at O as the y-axis, the equation of the path of P is

$$y = x - \frac{x^2}{500},$$

where x and y are measured in metres.

a By finding $\dfrac{dy}{dx}$, show that P was projected from O at an angle of 45° to the horizontal.

b Find the distance OA and the greatest vertical height attained by P above OA.

c Find the speed of projection of P.

d Find, to the nearest second, the time taken by P to move from O to A. **E**

8

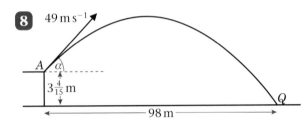

A golf ball is projected with speed $49\,\text{m s}^{-1}$ at an angle of elevation α from a point A on the first floor of a driving range. Point A is at a height of $3\frac{4}{15}\,\text{m}$ above horizontal ground. The ball first strikes the ground at a point Q which is at a horizontal distance of 98 m from the point A, as shown in the figure above.

a Show that

$$6\tan^2 \alpha - 30\tan \alpha + 5 = 0.$$

b Hence find the two possible angles of elevation.

c Find the smallest possible time of direct flight from A to Q. **E**

9 A particle P moves on the x-axis. At time t seconds, its acceleration is $(5 - 2t)\,\text{m s}^{-2}$, measured in the direction of x increasing. When $t = 0$, its velocity is $6\,\text{m s}^{-1}$ measured in the direction of x increasing. Find the time when P is instantaneously at rest in the subsequent motion. **E**

10 A particle P moves in a straight line in such a way that, at time t seconds, its velocity, $v\,\text{m s}^{-1}$, is given by

$$v = \begin{cases} 12t - 3t^2, & 0 \le t \le 5 \\ -\dfrac{375}{t^2}, & t > 5. \end{cases}$$

When $t = 0$, P is at the point O. Calculate the displacement of P from O

a when $t = 5$,

b when $t = 6$. **E**

11 A particle is moving in a straight line Ox. At time t seconds the acceleration of P is $a\,\text{m s}^{-2}$ and the velocity $v\,\text{m s}^{-1}$ of P is given by

$$v = 2 + 8\sin kt,$$

where k is a constant.

The initial acceleration of P is $4\,\text{m s}^{-2}$.

a Find the value of k.

Using the value of k found in **a**,

b find, in terms of π, the values of t in the interval $0 \le t \le 4\pi$ for which $a = 0$,

c show that $4a^2 = 64 - (v - 2)^2$. **E**

12 An aircraft is situated at rest at a point A on a runway XY which is of length 1400 m. Point A is 77 m from X. The aircraft moves along the runway towards Y with acceleration $\left(10 - \frac{4}{5}t\right)\,\text{m s}^{-2}$, where t seconds is the time from the instant the aircraft started to move.

a Find the speed of the aircraft when $t = 6$ and determine the distance travelled in the first 6 seconds of the aircraft's motion.

B is the point such that $AB = \frac{3}{10}AY$.

b Find the distance AB.

A safety regulation requires that the aircraft passes point B with a speed of $55\,\text{m s}^{-1}$ or more.

c Given that $t = T$ when the aircraft passes B, form an equation for T.

d Show that $T = 10.5$ satisfies the equation, and hence determine whether or not the aircraft satisfies this safety regulation as it passes B. **(E)**

13 A particle P moves along the x-axis. It passes through the origin O at time $t = 0$ with speed $7 \, \text{m s}^{-1}$ in the direction of x increasing.

At time t seconds the acceleration of P in the direction of x increasing is $(20 - 6t) \, \text{m s}^{-2}$.

a Show that the velocity $v \, \text{m s}^{-1}$ of P at time t seconds is given by
$$v = 7 + 20t - 3t^2.$$

b Show that $v = 0$ when $t = 7$ and find the greatest speed of P in the interval $0 \le t \le 7$.

c Find the distance travelled by P in the interval $0 \le t \le 7$. **(E)**

14 A particle P moves along a straight line. Initially, P is at rest at a point O on the line. At time t seconds (where $t \ge 0$) the acceleration of P is proportional to $(7 - t^2)$ and the displacement of P from O is s metres. When $t = 3$, the speed of P is $6 \, \text{m s}^{-1}$.

a Show that
$$s = \tfrac{1}{24} t^2 (42 - t^2).$$

b Find the total distance that P moves before returning to O. **(E)**

15 A particle P of mass $0.3 \, \text{kg}$ moves under the action of a single force $\mathbf{F}$ newtons. At time t seconds, the velocity $\mathbf{v} \, \text{m s}^{-1}$ of P is given by
$$\mathbf{v} = 3t^2 \mathbf{i} + (6t - 4)\mathbf{j}.$$

a Find the magnitude of $\mathbf{F}$ when $t = 2$.

When $t = 0$, P is at the point A. The position vector of A with respect to a fixed origin O is $(3\mathbf{i} - 4\mathbf{j}) \, \text{m}$. When $t = 4$, P is at the point B.

b Find the position vector of B. **(E)**

16 Referred to a fixed origin O, the position vector of a particle P at time t seconds is $\mathbf{r}$ metres, where
$$\mathbf{r} = 6t^2 \mathbf{i} + t^{\frac{5}{2}} \mathbf{j}, \; t \ge 0.$$

At the instant when $t = 4$, find

a the speed of P,

b the acceleration of P, giving your answer as a vector. **(E)**

17 A particle P moves in a horizontal plane. At time t seconds, the position vector of P is $\mathbf{r}$ metres relative to a fixed origin O where $\mathbf{r}$ is given by
$$\mathbf{r} = (18t - 4t^3)\mathbf{i} + ct^2 \mathbf{j},$$

where c is a positive constant. When $t = 1.5$, the speed of P is $15 \, \text{m s}^{-1}$. Find

a the value of c,

b the acceleration of P when $t = 1.5$. **(E)**

18 A particle P of mass $0.4 \, \text{kg}$ moves under the action of a single force $\mathbf{F}$ newtons. At time t seconds, the velocity of P, $\mathbf{v} \, \text{m s}^{-1}$, is given by
$$\mathbf{v} = (6t + 4)\mathbf{i} + (t^2 + 3t)\mathbf{j}.$$

When $t = 0$, P is at the point with position vector $(-3\mathbf{i} + 4\mathbf{j}) \, \text{m}$ with respect to a fixed origin O. When $t = 4$, P is at the point S.

a Calculate the magnitude of $\mathbf{F}$ when $t = 4$.

b Calculate the distance OS. **(E)**

19 Two particles P and Q move in a plane so that at time t seconds, where $t \geqslant 0$, P and Q have position vectors $\mathbf{r}_P$ metres and $\mathbf{r}_Q$ metres respectively, relative to a fixed origin O, where

$$\mathbf{r}_P = (3t^2 + 4)\mathbf{i} + \left(2t - \tfrac{1}{2}\right)\mathbf{j}$$

$$\mathbf{r}_Q = (t + 6)\mathbf{i} + \frac{3t^2}{2}\mathbf{j}.$$

Find

a the velocity vectors of P and Q at time t seconds,

b the speed of P when $t = 2$,

c the value of t at the instant when the particles are moving parallel to one another.

d Show that the particles collide and find the position vector of their point of collision. **(E)**

20 Referred to a fixed origin O, the particle R has position vector $\mathbf{r}$ metres at time t seconds, where

$$\mathbf{r} = (6 \sin \omega t)\mathbf{i} + (4 \cos \omega t)\mathbf{j}$$

and ω is a positive constant.

a Find $\dot{\mathbf{r}}$ and hence show that

$$v^2 = 2\omega^2 (13 + 5 \cos 2\omega t),$$

where $v\,\mathrm{m\,s}^{-1}$ is the speed of R at time t seconds.

b Deduce that

$$4\omega \leqslant v \leqslant 6\omega.$$

c Find $\ddot{\mathbf{r}}$.

d At the instant when $t = \dfrac{\pi}{3\omega}$, find the angle between $\dot{\mathbf{r}}$ and $\ddot{\mathbf{r}}$, giving your answer in degrees to one decimal place. **(E)**

21 Three particles of mass $3m$, $5m$ and λm are placed at the points with coordinates $(4, 0)$, $(0, -3)$ and $(4, 2)$ respectively.

The centre of mass of the three particles is at $(2, k)$.

a Show that $\lambda = 2$.

b Calculate the value of k. **(E)**

22 Particles of mass $2M$, xM and yM are placed at points whose coordinates are $(2, 5)$, $(1, 3)$ and $(3, 1)$ respectively. Given that the centre of mass of the three particles is at the point $(2, 4)$, find the values of x and y. **(E)**

23 Three particles of mass $0.1\,\mathrm{kg}$, $0.2\,\mathrm{kg}$ and $0.3\,\mathrm{kg}$ are placed at the points with position vectors $(2\mathbf{i} - \mathbf{j})\,\mathrm{m}$, $(2\mathbf{i} + 5\mathbf{j})\,\mathrm{m}$ and $(4\mathbf{i} + 2\mathbf{j})\,\mathrm{m}$ respectively. Find the position vector of the centre of mass of the particles. **(E)**

24 Three particles of mass $2M$, M and kM, where k is a constant, are placed at points with position vectors $6\mathbf{i}\,\mathrm{m}$, $4\mathbf{j}\,\mathrm{m}$ and $(2\mathbf{i} - 2\mathbf{j})\,\mathrm{m}$ respectively. The centre of mass of the three particles has position vector $(3\mathbf{i} + c\mathbf{j})\,\mathrm{m}$, where c is a constant.

a Show that $k = 3$.

b Hence find the value of c. **(E)**

25

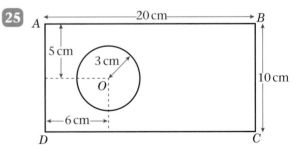

The figure shows a metal plate that is made by removing a circle of centre O and radius $3\,\mathrm{cm}$ from a uniform rectangular lamina $ABCD$, where $AB = 20\,\mathrm{cm}$ and $BC = 10\,\mathrm{cm}$. The point O is $5\,\mathrm{cm}$ from both AB and CD and is $6\,\mathrm{cm}$ from AD.

a Calculate, to 3 significant figures, the distance of the centre of mass of the plate from AD.

The plate is freely suspended from A and hangs in equilibrium.

b Calculate, to the nearest degree, the angle between AB and the vertical. **E**

---

26 A triangular frame ABC is made by bending a piece of wire of length 24 cm, so that AB, BC and AC are of lengths 6 cm, 8 cm and 10 cm respectively. Given that the wire is uniform, find the distance of the centre of mass of the frame from

a AB,

b BC.

The frame is suspended from the corner A and hangs in equilibrium.

c Find, to the nearest degree, the acute angle made by AB with the downward vertical. **E**

---

27 Three uniform rods AB, BC and CA of mass $2m$, m and $3m$ respectively have lengths l, l and $l\sqrt{2}$ respectively. The rods are rigidly joined to form a right-angled triangular framework.

a Calculate, in terms of l, the distance of the centre of mass of the framework from
  i BC,   **ii** AB.

b Calculate the angle, to the nearest degree, that BC makes with the vertical when the framework is freely suspended from the point B. **E**

28

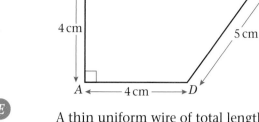

A thin uniform wire of total length 20 cm, is bent to form a frame. The frame is in the shape of a trapezium $ABCD$, where $AB = AD = 4$ cm, $CD = 5$ cm and AB is perpendicular to BC and AD, as shown in the figure.

a Find the distance of the centre of mass of the frame from AB.

The frame has mass M. A particle of mass kM is attached to the frame at C. When the frame is freely suspended from the mid-point of BC, the frame hangs in equilibrium with BC horizontal.

b Find the value of k. **E**

29

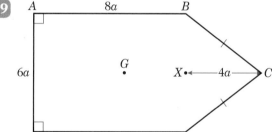

The figure shows a uniform lamina $ABCDE$ such that $ABDE$ is a rectangle, $BC = CD$, $AB = 8a$ and $AE = 6a$. The point X is the mid-point of BD and $XC = 4a$. The centre of mass of the lamina is at G.

a Show that $GX = \frac{44}{15}a$.

The mass of the lamina is M. A particle of mass λM is attached to the lamina at C. The lamina is suspended from B and hangs freely under gravity with AB horizontal.

b Find the value of λ. **E**

30 A uniform square plate *ABCD* has mass 10*M* and the length of a side of the plate is 2*l*. Particles of mass *M*, 2*M*, 3*M* and 4*M* are attached at *A*, *B*, *C* and *D* respectively. Calculate, in terms of *l*, the distance of the centre of mass of the loaded plate from

a *AB*,   **b** *BC*.

The loaded plate is freely suspended from the vertex *D* and hangs in equilibrium.

c Calculate, to the nearest degree, the angle made by *DA* with the downward vertical.   **E**

31

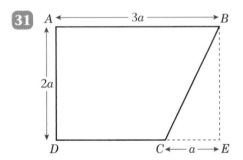

A uniform lamina *ABCD* is made by taking a uniform sheet of metal in the form of a rectangle *ABED*, with *AB* = 3*a* and *AD* = 2*a*, and removing the triangle *BCE*, where *C* lies on *DE* and *CE* = *a*, as shown in the figure.

a Find the distance of the centre of mass of the lamina from *AD*.

The lamina has mass *M*. A particle of mass *m* is attached to the lamina at *B*. When the loaded lamina is freely suspended from the mid-point of *AB*, it hangs in equilibrium with *AB* horizontal.

b Find *m* in terms of *M*.   **E**

32

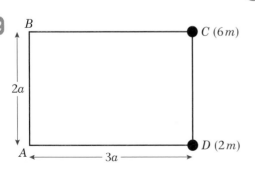

The figure shows four uniform rods joined to form a rectangular framework *ABCD*, where *AB* = *CD* = 2*a* and *BC* = *AD* = 3*a*. Each rod has mass *m*. Particles of mass 6*m* and 2*m* are attached to the framework at points *C* and *D* respectively.

a Find the distance of the centre of mass of the loaded framework from
   i *AB*,   **ii** *AD*.

The loaded framework is freely suspended from *B* and hangs in equilibrium.

b Find the angle which *BC* makes with the vertical.   **E**

33

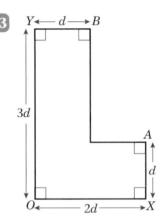

The figure shows a uniform L-shaped lamina with *OX* = 2*d*, *OY* = 3*d* and *AX* = *YB* = *d*. The angles at *O*, *A*, *B*, *X* and *Y* are all right angles.

Find, in terms of *d*, the distance of the centre of mass of the lamina

a from *OX*,   **b** from *OY*.

The lamina is suspended from the point *Y* and hangs freely in equilibrium.

c Find, to the nearest degree, the angle that *OY* makes with the vertical.   **E**

34

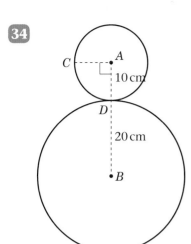

The figure shows a decoration which is made by cutting 2 circular discs from a sheet of uniform card. The discs are joined so that they touch at a point D on the circumference of both discs. The discs are coplanar and have centres A and B with radii 10 cm and 20 cm respectively.

a Find the distance of the centre of mass of the decoration from B.

The point C lies on the circumference of the smaller disc and $\angle CAB$ is a right angle. The decoration is freely suspended from C and hangs in equilibrium.

b Find, in degrees to one decimal place, the angle between AB and the vertical. **E**

35

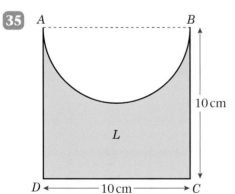

A uniform lamina L is formed by taking a uniform square sheet of material $ABCD$ of side 10 cm and removing a semicircle with diameter AB from the square, as shown in the figure.

a Find, in cm to 2 decimal places, the distance of the centre of mass of the lamina from the mid-point of AB.

[The centre of mass of a uniform semicircular lamina, radius a, is at a distance $\dfrac{4a}{3\pi}$ from the centre of the bounding diameter.]

The lamina is freely suspended from D and hangs at rest.

b Find, in degrees to one decimal place, the angle between CD and the vertical. **E**

36 A uniform lamina $ABCD$ is in the form of a trapezium in which $AB = AD = a$, $CD = 2a$ and $\angle BAD = \angle ADC = 90°$

a Find the distance of the centre of mass of the lamina from AD and from AB.

The lamina stands with the edge AB on a plane inclined at an angle α to the horizontal with A higher than B. The lamina is in a vertical plane through a line of greatest slope of the plane.

b Given that the lamina is on the point of overturning about B, find the value of $\tan \alpha$. **E**

37

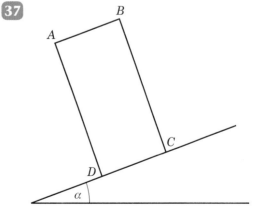

A thin uniform rectangular metal plate $ABCD$ of mass M rests on a rough plane inclined at an angle α to the horizontal. The plate lies in a vertical plane containing a line of greatest slope of the inclined plane, with the edge CD in contact with the plane and C further up the plane than D, as shown in the figure. The lengths of AB and BC are 10 cm and 30 cm respectively. The plane is sufficiently rough to prevent the plate from slipping.

a Find, to the nearest degree, the greatest value which α can have if the plate does not topple.

A small stud of mass m is fixed to the plate at the point C.

b Given that $\tan \alpha = \frac{1}{2}$, find, in terms of M, the smallest value of m which will enable the plate to stay in equilibrium without toppling. **E**

38

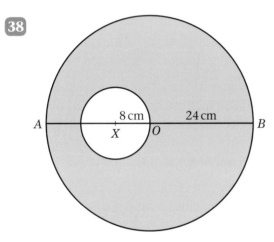

The figure shows a template T made by removing a circular disc, of centre X and radius 8 cm, from a uniform circular lamina, of centre O and radius 24 cm. The point X lies on the diameter AOB of the lamina and $AX = 16$ cm. The centre of mass of T lies at the point G.

a Find AG.

The template T is free to rotate about a smooth fixed horizontal axis, perpendicular to the plane of T, which passes through the mid-point of OB. A small stud of mass $\frac{1}{4}m$ is fixed at B, and T and the stud are in equilibrium with AB horizontal.

b Modelling the stud as a particle, find the mass of T in terms of m. **E**

39

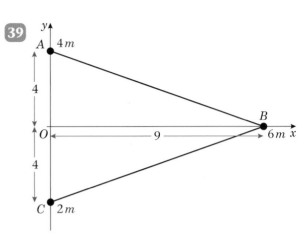

The figure shows a triangular lamina ABC. The coordinates of A, B and C are (0, 4), (9, 0) and (0, −4) respectively. Particles of mass 4m, 6m and 2m are attached at A, B and C respectively.

a Calculate the coordinates of the centre of mass of the three particles, *without the lamina*.

The lamina ABC is uniform and of mass km. The centre of mass of the combined system consisting of the three particles and the lamina has coordinates $(4, \lambda)$.

b Show that $k = 6$.

c Calculate the value of λ.

The combined system is freely suspended from O and hangs at rest.

d Calculate, in degrees to one decimal place, the angle between AC and the vertical. **E**

40

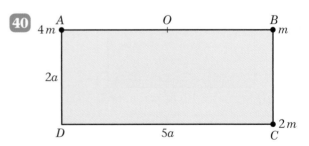

A loaded plate L is modelled as a uniform rectangular lamina ABCD and three particles. The sides CD and AD of the lamina have length 5*a* and 2*a* respectively and the mass of the lamina is 3*m*. The three particles have mass 4*m*, *m* and 2*m* and are attached at the points *A*, *B* and *C* respectively, as shown in the figure.

a Show that the distance of the centre of mass of *L* from *AD* is 2.25*a*.

b Find the distance of the centre of mass of *L* from *AB*.

The point *O* is the mid-point of *AB*. The loaded plate *L* is freely suspended from *O* and hangs at rest under gravity.

c Find, to the nearest degree, the size of the angle that *AB* makes with the horizontal.

A horizontal force of magnitude *P* is applied at *C* in the direction *CD*. The loaded plate *L* remains suspended from *O* and rests in equilibrium with *AB* horizontal and *C* vertically below *B*.

d Show that $P = \frac{5}{4}mg$.

e Find the magnitude of the force on *L* at *O*. **E**

41 Two particles, of mass 3*m* and 2*m*, are moving in opposite directions in a straight horizontal line with speeds 4*u* and *u* respectively. The particles collide and coalesce to form a single particle *P*. Calculate

a the speed of *P* in terms of *u*,

b the loss in kinetic energy, in terms of *m* and *u*, due to the collision. **E**

42 A particle of mass 4 kg is moving in a straight horizontal line. There is a constant resistive force of magnitude *R* newtons. The speed of the particle is reduced from 25 m s⁻¹ to rest over a distance of 200 m.

Use the work–energy principle to calculate the value of *R*. **E**

43

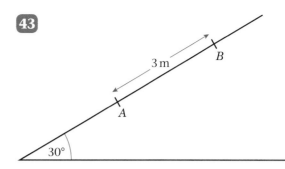

A particle *P* of mass 2 kg is projected from a point *A* up a line of greatest slope *AB* of a fixed plane. The plane is inclined at an angle of 30° to the horizontal and *AB* = 3 m with *B* above *A*, as shown in the figure. The speed of *P* at *A* is 10 m s⁻¹.

a Assuming the plane is smooth, find the speed of *P* at *B*.

The plane is now assumed to be rough. At *A* the speed of *P* is 10 m s⁻¹ and at *B* the speed of *P* is 7 m s⁻¹.

b By using the work–energy principle, or otherwise, find the coefficient of friction between *P* and the plane. **E**

44

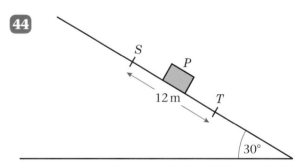

A small package is modelled as a particle *P* of mass 0.6 kg. The package slides down a rough plane from a point *S* to a point *T*, where *ST* = 12 m. The plane is inclined at 30° to the horizontal and *ST* is a line of greatest slope of the plane, as shown in the figure. The speed of *P* at *S* is 10 m s⁻¹ and the speed of *P* at *T* is 9 m s⁻¹. Calculate

a the total loss of energy of *P* in moving from *S* to *T*,

b the coefficient of friction between *P* and the plane. **E**

45 A particle P has mass 4 kg. It is projected from a point A up a line of greatest slope of a rough plane inclined at an angle α to the horizontal, where $\tan \alpha = \frac{3}{4}$. The coefficient of friction between P and the plane is $\frac{2}{7}$. The particle comes to rest instantaneously at the point B on the plane, where $AB = 2.5$ m. It then moves back down the plane to A.

a Find the work done by friction as P moves from A to B.

b Using the work–energy principle, find the speed with which P is projected from A.

c Find the speed of P when it returns to A. **E**

46

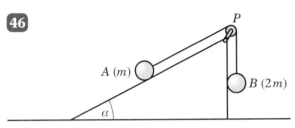

Two particles A and B of mass m and $2m$ respectively are attached to the ends of a light inextensible string. The particle A lies on a rough plane inclined at an angle α to the horizontal, where $\tan \alpha = \frac{3}{4}$. The string passes over a small light pulley P fixed at the top of the plane. The particle B hangs freely below P, as shown in the figure. The particles are released from rest with the string taut and the section of the string from A to P parallel to a line of greatest slope of the plane. The coefficient of friction between A and the plane is $\frac{5}{8}$. When each particle has moved a distance h, B has not reached the ground and A has not reached P.

a Find an expression for the potential energy lost by the system when each particle has moved a distance h.

When each particle has moved a distance h, they are moving with speed v.

b Using the work–energy principle, find an expression for v^2, giving your answer in the form kgh where k is a number. **E**

47 The engine of a car is working at a constant rate of 6 kW in driving a car along a straight horizontal road at 54 km h^{-1}. Find, in N, the magnitude of the resistance to motion of the car. **E**

48 A lorry of mass 5000 kg moves at a constant speed of 15 m s^{-1} up a hill inclined at an angle α to the horizontal, where $\sin \alpha = \frac{1}{15}$. The resistance experienced by the lorry is constant and has magnitude 2500 N.

Find, in kW, the rate of working of the lorry's engine. **E**

49 A car of mass 400 kg is moving up a straight road inclined at an angle θ to the horizontal where $\sin \theta = \frac{1}{14}$. The resistance to motion of the car from non-gravitational forces is modelled as a constant force of magnitude R newtons. When the car is moving at a constant speed of 20 m s^{-1}, the power developed by the car's engine is 10 kW.

Find the value of R. **E**

50 A lorry of mass 1500 kg moves along a straight horizontal road. The resistance to motion of the lorry has magnitude 750 N and the lorry's engine is working at a rate of 36 kW.

a Find the acceleration of the lorry when its speed is 20 m s^{-1}.

The lorry comes to a hill inclined at an angle α to the horizontal, where $\sin \alpha = \frac{1}{10}$. The magnitude of the resistance to motion from non-gravitational forces remains 750 N.

The lorry moves up the hill at a constant speed of $20\,\text{m s}^{-1}$.

b Find the rate at which the lorry is now working.

51 A car of mass $1200\,\text{kg}$ moves along a straight horizontal road. The resistance to motion of the car from non-gravitational forces is of constant magnitude $600\,\text{N}$. The car moves with constant speed and the engine of the car is working at a rate of $21\,\text{kW}$.

a Find the speed of the car.

The car moves up a hill inclined at an angle α to the horizontal, where $\sin \alpha = \frac{1}{14}$.

The car's engine continues to work at $21\,\text{kW}$ and the resistance to motion from non-gravitational forces remains of magnitude $600\,\text{N}$.

b Find the constant speed at which the car moves up the hill. *E*

52 A car of mass $800\,\text{kg}$ tows a caravan of mass $480\,\text{kg}$ along a straight level road. The tow-bar connecting the car and the caravan is horizontal and of negligible mass. With the car's engine working at a rate of $24\,\text{kW}$, the car and caravan are travelling at a constant speed of $25\,\text{m s}^{-1}$.

a Calculate the magnitude of the total resistance to the motion of the car and the caravan.

The resistance to the motion of the car has magnitude 800λ newtons and the resistance to the motion of the caravan has magnitude 480λ newtons, where λ is a constant. Find

b the value of λ,

c the tension in the tow-bar. *E*

53 A car of mass $1000\,\text{kg}$ is moving along a straight horizontal road. The resistance to motion is modelled as a constant force

of magnitude R newtons. The engine of the car is working at a constant rate of $12\,\text{kW}$. When the car is moving with speed $15\,\text{m s}^{-1}$, the acceleration of the car is $0.2\,\text{m s}^{-2}$.

a Show that $R = 600$.

The car now moves with constant speed $U\,\text{m s}^{-1}$ downhill on a straight road inclined at θ to the horizontal, where $\sin \theta = \frac{1}{40}$. The engine of the car is now working at a rate of $7\,\text{kW}$. The resistance to motion from non-gravitational forces remains of magnitude R newtons.

b Calculate the value of U. *E*

54 A car of mass $1000\,\text{kg}$ is moving along a straight road with constant acceleration $f\,\text{m s}^{-2}$. The resistance to motion is modelled as a constant force of magnitude $1200\,\text{N}$. When the car is travelling at $12\,\text{m s}^{-1}$, the power generated by the engine of the car is $24\,\text{kW}$.

a Calculate the value of f.

When the car is travelling at $14\,\text{m s}^{-1}$, the engine is switched off and the car comes to rest without braking in a distance d metres.

b Assuming the same model for resistance, use the work–energy principle to calculate the value of d.

c Give a reason why the model used for resistance may not be realistic. *E*

55

The figure shows the path taken by a cyclist in travelling on a section of a road. When the cyclist comes to the point A on the top of the hill she is travelling at $8\,\text{m s}^{-1}$. She descends a vertical distance of

20 m to the bottom of the hill. The road then rises to the point B through a vertical distance of 12 m. When she reaches B her speed is $5\,\mathrm{m\,s^{-1}}$. The total mass of the cyclist and the cycle is 80 kg and the total distance along the road from A to B is 500 m. By modelling the resistance to the motion of the cyclist as of constant magnitude 20 N,

a find the work done by the cyclist in moving from A to B.

At B the road is horizontal.

b Given that at B the cyclist is accelerating at $0.5\,\mathrm{m\,s^{-2}}$, find the power generated by the cyclist at B. **(E)**

56 A van of mass 1500 kg is driving up a straight road inclined at an angle α to the horizontal, where $\sin\alpha = \frac{1}{12}$. The resistance to motion due to non-gravitational forces is modelled as a constant force of magnitude 1000 N.

a Given that initially the speed of the van is $30\,\mathrm{m\,s^{-1}}$ and that the van's engine is working at a rate of 60 kW, calculate the magnitude of the initial deceleration of the van.

When travelling up the same hill, the rate of working of the van's engine is increased to 80 kW.

b Using the same model for the resistance due to non-gravitational forces, calculate in $\mathrm{m\,s^{-1}}$ the constant speed which can be sustained by the van at this rate of working.

c Give one reason why the use of this model for resistance may mean your answer to part **b** is too high. **(E)**

57 A model car has weight 200 N. It undergoes tests on a straight hill inclined at an angle α to the horizontal, where $\sin\alpha = \frac{1}{10}$. The engine of the car works at a constant rate of P watts.

When the car goes up the hill it is observed to travel at a constant speed of $8\,\mathrm{m\,s^{-1}}$. Given that the total resistance to the motion of the car from forces other than gravity is R newtons,

a express P in terms of R.

When the car runs down the same hill with the engine running at the same rate, it is observed to travel at a constant speed of $24\,\mathrm{m\,s^{-1}}$.

In an initial model of the situation the resistance to motion due to non-gravitational forces is assumed to be constant whatever the speed of the car.

b Using this model, find an estimate for the value of P.

In a refined model the resistance to motion due to non-gravitational forces is assumed to be proportional to the speed of the car.

c Using this model, find a revised estimate for P.

58 A car of mass 1000 kg is towing a trailer of mass 1500 kg along a straight horizontal road. The tow-bar joining the car to the trailer is modelled as a light rod parallel to the road. The total resistance to motion of the car is modelled as having constant magnitude 750 N. The total resistance to motion of the trailer is modelled as a force of magnitude R newtons, where R is a constant. When the engine is working at a rate of 50 kW, the car and the trailer travel at a constant speed of $25\,\mathrm{m\,s^{-1}}$.

a Show that $R = 1250$.

When travelling at $25\,\mathrm{m\,s^{-1}}$ the driver of the car disengages the engine and applies the brakes. The brakes provide a constant braking force of magnitude 1500 N to the car. The resisting forces of magnitude 750 N and 1250 N are assumed to remain unchanged. Calculate

b the deceleration of the car while braking,

c the thrust in the tow-bar while braking,

d the work done, in kJ, by the braking force in bringing the car and the trailer to rest.

e Suggest how the modelling assumption that the resistances to motion are constant could be refined to be more realistic. **E**

59

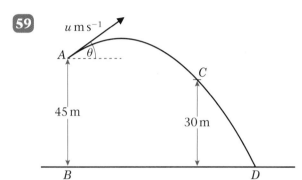

A particle P is projected from a point A with speed $u\,\mathrm{m\,s^{-1}}$ at an angle of elevation θ, where $\cos\theta = \frac{4}{5}$. The point B, on horizontal ground, is vertically below A and $AB = 45$ m. After projection, P moves freely under gravity passing through a point C, 30 m above the horizontal ground, before striking the ground at the point D, as shown in the figure above.

Given that P passes through C with speed $24.5\,\mathrm{m\,s^{-1}}$,

a using conservation of energy, or otherwise, show that $u = 17.5$,

b find the size of the angle which the velocity of P makes with the horizontal as P passes through C,

c find the distance BD.

60

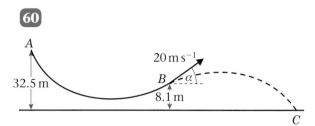

In a ski-jump competition, a skier of mass 80 kg moves from rest at a point A on a ski-slope. The skier's path is an arc AB. The starting point A of the slope is 32.5 m above horizontal ground. The end B of the slope is 8.1 m above the ground. When the skier reaches B she is travelling at $20\,\mathrm{m\,s^{-1}}$ and moving upwards at an angle α to the horizontal, where $\tan\alpha = \frac{3}{4}$, as shown in the figure. The distance along the slope from A to B is 60 m. The resistance to motion while she is on the slope is modelled as a force of constant magnitude R newtons.

a By using the work–energy principle, find the value of R.

On reaching B, the skier then moves through the air and reaches the ground at the point C. The motion of the skier in moving from B to C is modelled as that of a particle moving freely under gravity.

b Find the time the skier takes to move from B to C.

c Find the horizontal distance from B to C.

d Find the speed of the skier immediately before she reaches C. **E**

After completing this chapter you should be able to

1 use the impulse–momentum principle and the principle of conservation of linear momentum in vector form

2 apply conservation of linear momentum and Newton's Law of Restitution to solve problems involving direct impact

3 model and solve problems involving successive impacts.

Collisions

There are a number of situations which can be modelled as impulse–momentum problems and which can be tackled using the equations and formulae that you will meet in this chapter.

Examples include a bat hitting a ball, a snooker ball hitting another snooker ball, a jerk in a string when the string suddenly goes tight, the recoil of a rifle when a shot is fired and a ball hitting a surface at right angles.

This chapter builds on the work on collisions covered in chapter 3 of book M1.

4.1 You can use the impulse–momentum principle and the principle of conservation of linear momentum in vector form.

In M1 you used the following impulse–momentum principle and the principle of conservation of linear momentum.

The **impulse–momentum principle** states that the impulse of a force is equal to the change in momentum produced:

impulse = force × time
$$I = mv - mu$$
Impulse is measured in Newton-seconds (N s).

The **principle of conservation of linear momentum** states that the total momentum before impact equals the total momentum after impact:

momentum = mass × velocity
$$m_1u_1 + m_2u_2 = m_1v_1 + m_2v_2$$

Momentum is measured in $kg\,m\,s^{-1}$.

Impulse and momentum are both vector quantities. You can write the impulse–momentum principle and the principle of conservation of linear momentum as vector equations, and use them to solve problems involving collisions where the velocities and any impulse are given in vector form.

- $\mathbf{I} = m\mathbf{v} - m\mathbf{u}$
  where m is the mass of the body, $\mathbf{u}$ the initial velocity and $\mathbf{v}$ the final velocity.

- $m_1\mathbf{u}_1 + m_2\mathbf{u}_2 = m_1\mathbf{v}_1 + m_2\mathbf{v}_2$
  where a body of mass m_1 moving with velocity $\mathbf{u}_1$ collides with a body of mass m_2 moving with a velocity of $\mathbf{u}_2$. $\mathbf{v}_1$ and $\mathbf{v}_2$ are the velocities of the bodies after the collision.

Example 1

A particle of mass 0.2 kg is moving with velocity $(10\mathbf{i} - 5\mathbf{j})\,m\,s^{-1}$ when it receives an impulse $(3\mathbf{i} - 2\mathbf{j})$ Ns. Find the new velocity of the particle.

The change in momentum of the particle is
$$0.2\mathbf{v} - 0.2(10\mathbf{i} - 5\mathbf{j})\,kg\,m\,s^{-1}$$

Let the velocity of the particle after the impact be $\mathbf{v}\,m\,s^{-1}$.

From the impulse–momentum principle this is equal to the impulse:
$$0.2\mathbf{v} - 0.2(10\mathbf{i} - 5\mathbf{j}) = 3\mathbf{i} - 2\mathbf{j}$$

Use $m\mathbf{v} - m\mathbf{u} = \mathbf{I}$ substituting $m = 0.2$, $\mathbf{u} = (10\mathbf{i} - 5\mathbf{j})$ and $\mathbf{I} = 3\mathbf{i} - 2\mathbf{j}$

$$0.2\mathbf{v} = 3\mathbf{i} - 2\mathbf{j} + 2\mathbf{i} - \mathbf{j}$$
$$= 5\mathbf{i} - 3\mathbf{j}$$
$$\mathbf{v} = 25\mathbf{i} - 15\mathbf{j}$$

Make $\mathbf{v}$ the subject.

Example 2

An ice hockey puck of mass 0.17 kg receives an impulse of $\mathbf{Q}$ Ns. Immediately before the impulse the velocity of the puck is $(10\mathbf{i} + 5\mathbf{j})\,\mathrm{m\,s^{-1}}$ and immediately afterwards its velocity is $(15\mathbf{i} - 7\mathbf{j})\,\mathrm{m\,s^{-1}}$. Find the magnitude of $\mathbf{Q}$ and the angle between $\mathbf{Q}$ and $\mathbf{i}$.

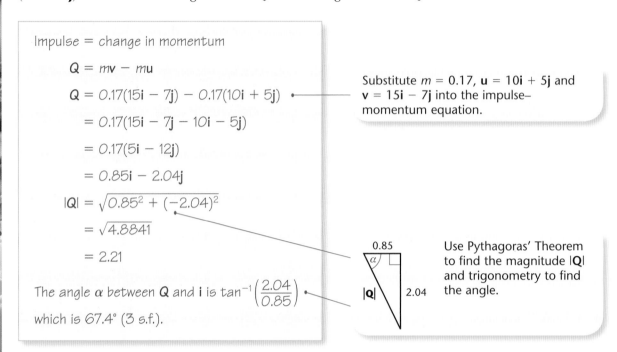

Impulse = change in momentum

$\mathbf{Q} = m\mathbf{v} - m\mathbf{u}$

$\mathbf{Q} = 0.17(15\mathbf{i} - 7\mathbf{j}) - 0.17(10\mathbf{i} + 5\mathbf{j})$ Substitute $m = 0.17$, $\mathbf{u} = 10\mathbf{i} + 5\mathbf{j}$ and $\mathbf{v} = 15\mathbf{i} - 7\mathbf{j}$ into the impulse–momentum equation.

$\quad = 0.17(15\mathbf{i} - 7\mathbf{j} - 10\mathbf{i} - 5\mathbf{j})$

$\quad = 0.17(5\mathbf{i} - 12\mathbf{j})$

$\quad = 0.85\mathbf{i} - 2.04\mathbf{j}$

$|\mathbf{Q}| = \sqrt{0.85^2 + (-2.04)^2}$

$\quad = \sqrt{4.8841}$

$\quad = 2.21$

Use Pythagoras' Theorem to find the magnitude $|\mathbf{Q}|$ and trigonometry to find the angle.

The angle α between $\mathbf{Q}$ and $\mathbf{i}$ is $\tan^{-1}\left(\dfrac{2.04}{0.85}\right)$ which is $67.4°$ (3 s.f.).

Example 3

A squash ball of mass 0.025 kg is moving with velocity $(22\mathbf{i} + 37\mathbf{j})\,\mathrm{m\,s^{-1}}$ when it hits a wall. It rebounds with velocity $(10\mathbf{i} - 11\mathbf{j})\,\mathrm{m\,s^{-1}}$. Find the impulse exerted by the wall on the squash ball.

Impulse $= m\mathbf{v} - m\mathbf{u}$

Impulse $= 0.025((10\mathbf{i} - 11\mathbf{j}) - (22\mathbf{i} + 37\mathbf{j}))$ The impulse exerted by the wall on the squash ball is equal to the change in momentum of the ball.

$\quad = 0.025(-12\mathbf{i} - 48\mathbf{j})$

$\quad = (-0.3\mathbf{i} - 1.2\mathbf{j})\,\mathrm{Ns}$

Example 4

A particle of mass 0.15 kg is moving with velocity $(20\mathbf{i} - 10\mathbf{j})\,\mathrm{m\,s^{-1}}$ when it collides with a particle of mass 0.25 kg moving with velocity $(16\mathbf{i} - 8\mathbf{j})\,\mathrm{m\,s^{-1}}$. The two particles coalesce and form one particle of mass 0.4 kg. Find the velocity of the combined particle.

$m_1u_1 + m_2u_2 = m_1v_1 + m_2v_2$

$0.15(20i - 10j) + 0.25(16i - 8j) = 0.4v$

$3i - 1.5j + 4i - 2j = 0.4v$

$7i - 3.5j = 0.4v$

$v = 17.5i - 8.75j$

The velocity of the combined particle is $(17.5i - 8.75j) \, m\,s^{-1}$.

This is the vector form of the conservation of linear momentum equation.

After the impact $v_1 = v_2 = v$ and the equation becomes $m_1v_1 + m_2v_2 = (m_1 + m_2)v$.

Note that the velocity vectors are all parallel and the question involves direct impact. Oblique impact is covered in a later unit.

Exercise 4A

In this exercise **i** and **j** are perpendicular unit vectors.

1. A particle of mass 0.25 kg is moving with velocity $(12i + 4j) \, m\,s^{-1}$ when it receives an impulse $(8i - 7j) \, Ns$. Find the new velocity of the particle.

2. A particle of mass 0.5 kg is moving with velocity $(2i - 2j) \, m\,s^{-1}$ when it receives an impulse $(3i + 5j) \, Ns$. Find the new velocity of the particle.

3. A particle of mass 2 kg moves with velocity $(3i + 2j) \, m\,s^{-1}$ immediately after it has received an impulse $(4i + 8j) \, Ns$. Find the original velocity of the particle.

4. A particle of mass 1.5 kg moves with velocity $(5i - 8j) \, m\,s^{-1}$ immediately after it has received an impulse $(3i - 6j) \, Ns$. Find the original velocity of the particle.

5. A body of mass 3 kg is initially moving with a constant velocity of $(i + j) \, m\,s^{-1}$ when it is acted on by a force of $(6i - 8j) \, N$ for 3 seconds. Find the impulse exerted on the body and find its velocity when the force ceases to act.

6. A body of mass 0.5 kg is initially moving with a constant velocity of $(5i + 12j) \, m\,s^{-1}$ when it is acted on by a force of $(2i - j) \, N$ for 5 seconds. Find the impulse exerted on the body and find its velocity when the force ceases to act.

7. A particle of mass 2 kg is moving with velocity $(5i + 3j) \, m\,s^{-1}$ when it hits a wall. It rebounds with velocity $(-i - 3j) \, m\,s^{-1}$. Find the impulse exerted by the wall on the particle.

8. A particle of mass 0.5 kg is moving with velocity $(11i - 2j) \, m\,s^{-1}$ when it hits a wall. It rebounds with velocity $(-i + 7j) \, m\,s^{-1}$. Find the impulse exerted by the wall on the particle.

9. A particle P of mass 3 kg receives an impulse **Q** Ns. Immediately before the impulse the velocity of P is $5i \, m\,s^{-1}$ and immediately afterwards it is $(13i - 6j) \, m\,s^{-1}$. Find the magnitude of **Q** and the angle between **Q** and **i**.

10 A particle P of mass 0.5 kg receives an impulse $\mathbf{Q}$ Ns. Immediately before the impulse the velocity of P is $(-\mathbf{i} - 2\mathbf{j})$ m s^{-1} and immediately afterwards it is $(3\mathbf{i} - 4\mathbf{j})$ m s^{-1}. Find the magnitude of $\mathbf{Q}$ and the angle between $\mathbf{Q}$ and $\mathbf{i}$.

11 A cricket ball of mass 0.5 kg is hit by a bat. Immediately before being hit the velocity of the ball is $(20\mathbf{i} - 4\mathbf{j})$ m s^{-1} and immediately afterwards it is $(-16\mathbf{i} + 8\mathbf{j})$ m s^{-1}. Find the magnitude of the impulse exerted on the ball by the bat.

12 A ball of mass 0.2 kg is hit by a bat. Immediately before being hit by the bat the velocity of the ball is $-15\mathbf{i}$ m s^{-1} and the bat exerts an impulse of $(2\mathbf{i} + 6\mathbf{j})$ Ns on the ball. Find the velocity of the ball after the impact.

13 A particle of mass 0.25 kg has velocity $\mathbf{v}$ m s^{-1} at time t s where $\mathbf{v} = (t^2 - 3)\mathbf{i} + 4t\mathbf{j}$. When $t = 3$, the particle receives an impulse of $(2\mathbf{i} + 2\mathbf{j})$ Ns. Find the velocity of the particle immediately after the impulse.

14 A ball of mass 2 kg is initially moving with a velocity of $(\mathbf{i} + \mathbf{j})$ m s^{-1}. It receives an impulse of $2\mathbf{j}$ Ns. Find the velocity immediately after the impulse and the angle through which the ball is deflected as a result. Give your answer to the nearest degree.

15 A particle of mass 0.5 kg moving with velocity $3\mathbf{i}$ m s^{-1} collides with a particle of mass 0.25 kg moving with velocity $12\mathbf{i}$ m s^{-1}. The two particles coalesce and move as one particle of mass 0.75 kg. Find the velocity of the combined particle.

16 A particle of mass 5 kg moving with velocity $(\mathbf{i} - \mathbf{j})$ m s^{-1} collides with a particle of mass 2 kg moving with velocity $(-\mathbf{i} + \mathbf{j})$ m s^{-1}. The two particles coalesce and move as one particle of mass 7 kg. Find the magnitude of the velocity $\mathbf{v}$ m s^{-1} of the combined particle.

4.2 You can solve problems involving the direct impact of two particles by using conservation of linear momentum and Newton's Law of Restitution.

A direct impact is a collision between particles, of the same size, which are moving along the same straight line. When two particles collide their speeds after the collision depend upon the material from which they are made.

Newton's Law of Restitution (sometimes called Newton's Experimental Law) defines how the speeds of the particles after the collision depend on the nature of the particles as well as the speeds before the collision. This law only holds when the collision takes place in free space or on a smooth surface.

■ $\dfrac{\textbf{speed of separation of particles}}{\textbf{speed of approach of particles}} = e$

The constant e is called the coefficient of restitution and $0 \leqslant e \leqslant 1$.

The value of the coefficient of restitution e depends on the materials from which the particles are made. Particles for which $e = 1$ are called **perfectly elastic** particles. Particles for which $e = 0$ are called **inelastic particles**. Inelastic particles coalesce on impact.

Table tennis balls and some marbles are very elastic with values for e of approximately 0.95. Tennis balls, golf balls and cricket balls are less elastic with values for e ranging between 0.4 and 0.9. Balls of plasticine are inelastic and e is zero.

	Elastic particles ($e = 1$)	Inelastic particles ($e = 0$)
Before collision:	$v\,\mathrm{m\,s^{-1}}$ → ← $v\,\mathrm{m\,s^{-1}}$ ○ ○	$v\,\mathrm{m\,s^{-1}}$ → ← $v\,\mathrm{m\,s^{-1}}$ ○ ○
After collision:	$v\,\mathrm{m\,s^{-1}}$ ← → $v\,\mathrm{m\,s^{-1}}$ ○ ○	$0\,\mathrm{m\,s^{-1}}$ ∞

Example 5

In each part of this question the two diagrams show the speeds of two particles A and B just before and just after a collision. The particles move on a smooth horizontal plane. Find the coefficient of restitution e in each case.

a

Before impact	After impact
$8\,\mathrm{m\,s^{-1}}$ → At rest ○ ○ A B	At rest $2\,\mathrm{m\,s^{-1}}$ → ○ ○ A B

a The speed of approach is $8 - 0 = 8\,\mathrm{m\,s^{-1}}$

The speed of separation is $2 - 0 = 2\,\mathrm{m\,s^{-1}}$

The value of e is $\frac{2}{8} = \frac{1}{4}$

You should find the difference in the speeds before impact, called the speed of approach. This must be positive for the collision to take place.

You should also find the difference in the speeds after impact, called the speed of separation. This must be positive for separation to take place after the collision.

Find e using
$$\frac{\text{speed of separation of particles}}{\text{speed of approach of particles}} = e.$$

b

Before impact	After impact
6 m s⁻¹ → Ⓐ 3 m s⁻¹ → Ⓑ	4 m s⁻¹ → Ⓐ 5 m s⁻¹ → Ⓑ

b The speed of approach is $6 - 3 = 3 \, \text{m s}^{-1}$

The speed of separation is $5 - 4 = 1 \, \text{m s}^{-1}$

The value of e is $\frac{1}{3}$.

Find e using

$$\frac{\text{speed of separation of particles}}{\text{speed of approach of particles}} = e$$

c

Before impact	After impact
11 m s⁻¹ → Ⓐ 7 m s⁻¹ ← Ⓑ	6 m s⁻¹ ← Ⓐ 3 m s⁻¹ → Ⓑ
A B	A B

The speed of B before the collision is negative as B is moving to the left.

c The speed of approach is $11 - (-7) = 11 + 7 = 18 \, \text{m s}^{-1}$

The speed of separation is $3 - (-6) = 3 + 6 = 9 \, \text{m s}^{-1}$

The value of e is $\frac{9}{18} = \frac{1}{2}$

The speed of A after the collision is negative as A is moving to the left.

Example 6

Find the value of v in the situation shown, given that the coefficient of restitution e is $\frac{1}{3}$.

Before collision	After collision
4 m s⁻¹ → Ⓐ 3 m s⁻¹ → Ⓑ	2 m s⁻¹ → Ⓐ v m s⁻¹ → Ⓑ
A B	A B

$$\frac{\text{speed of separation of particles}}{\text{speed of approach of particles}} = e$$

$$\frac{v - 2}{4 - 3} = \frac{1}{3}$$

$$\therefore v - 2 = \frac{1}{3}$$

So $v = 2\frac{1}{3}$

Substitute the speed of approach, $4 - 3$, and the speed of separation, $v - 2$, then make v the subject of the formula.

■ You can use the principle of conservation of linear momentum together with Newton's Law of Restitution to solve problems involving two unknown velocities.

Example 7

Find the values of v_1 and v_2 in the situation shown, given that the coefficient of restitution e is $\frac{1}{2}$.

Before collision	After collision	Before collision	After collision
$5\,\mathrm{m\,s^{-1}}$ →	$4\,\mathrm{m\,s^{-1}}$ ←	$v_1\,\mathrm{m\,s^{-1}}$ →	$v_2\,\mathrm{m\,s^{-1}}$ →
$A\,(200\,\mathrm{g})$	$B\,(400\,\mathrm{g})$	$A\,(200\,\mathrm{g})$	$B\,(400\,\mathrm{g})$

Using conservation of linear momentum for the system ($\rightarrow$)

$$0.2 \times 5 + 0.4 \times (-4) = 0.2v_1 + 0.4v_2$$
$$1 - 1.6 = 0.2v_1 + 0.4v_2$$
$$-0.6 = 0.2v_1 + 0.4v_2$$
$$-3 = v_1 + 2v_2 \quad \textbf{①}$$

Use $m_1\mathbf{u}_1 + m_2\mathbf{u}_2 = m_1\mathbf{v}_1 + m_2\mathbf{v}_2$, noting that u_2 is negative as B is moving to the left.

$$\frac{\text{speed of separation of particles}}{\text{speed of approach of particles}} = e$$

$$\frac{v_2 - v_1}{5 - (-4)} = \frac{1}{2}$$

$$v_2 - v_1 = 4\tfrac{1}{2} \quad \textbf{②}$$

Calculate the speed of approach and the speed of separation and substitute into Newton's Law of Restitution.

Eliminating v_1 between equations ① and ② gives

$$v_2 = \frac{1}{2}$$

Substituting this value into equation ① gives

$$v_1 = -4$$

Solve the simultaneous equations ① and ② to find the values of v_1 and v_2.

Example 8

Two small spheres P and Q have mass $3m$ and $4m$ respectively. They are moving towards each other in opposite directions on a smooth horizontal plane. P has speed $3u$ and Q has speed $2u$ just before the impact. The coefficient of restitution between P and Q is e.

a Show that the speed of Q after the collision is $\frac{u}{7}(15e + 1)$.

b Given that the direction of motion of P is unchanged, find the range of possible values of e.

c Given that the magnitude of the impulse of P on Q is $\frac{80mu}{9}$, find the value of e.

a

Before impact		After impact	
$3u$	$2u$	v_1	v_2
→ ◯	← ◯	→ ◯	→ ◯
$P\,(3m)$	$Q\,(4m)$	$P\,(3m)$	$Q\,(4m)$

Draw a diagram showing the masses and speeds before and after the impact. Use v_1 and v_2 for unknown speeds after impact.

Using conservation of linear momentum for the system (→):

$$9mu - 8mu = 3mv_1 + 4mv_2$$

$$\Rightarrow 3v_1 + 4v_2 = u \qquad \textbf{①}$$

Use $m_1\mathbf{u}_1 + m_2\mathbf{u}_2 = m_1\mathbf{v}_1 + m_2\mathbf{v}_2$, noting that u_2 is negative as Q is moving to the left.

Newton's Law of Restitution gives

$$\frac{v_2 - v_1}{3u + 2u} = e$$

$$\Rightarrow v_2 - v_1 = 5eu \qquad \textbf{②}$$

Calculate the speed of approach $3u - (-2u)$ and the speed of separation $v_2 - v_1$ and substitute into Newton's Law of Restitution.

Eliminating v_1 between equations **①** and **②** gives

$$7v_2 = 15ue + u$$

$$v_2 = \frac{u}{7}(15e + 1)$$

Solve the simultaneous equations **①** and **②** to find the value of v_2.

b Substituting this value into equation **②** gives

$$v_1 = \frac{u}{7}(15e + 1) - 5eu$$

$$v_1 = \frac{u}{7}(1 - 20e)$$

Now find the value of v_1 by substituting the value of v_2 from **a** into equation **②**.

As $v_1 > 0$,

$$\frac{u}{7}(1 - 20e) > 0$$

So $e < \frac{1}{20}$.

As the direction of motion of P is unchanged by the impact, P continues to move to the right and you need to **use the condition that $v_1 > 0$.**

c Impulse of P on Q = change in momentum of Q

$$= 4mv_2 - 4m(-2u)$$

$$= \frac{4mu}{7}(15e + 1) + 8mu$$

$$= \frac{60mu}{7}(1 + e)$$

This is the impulse–momentum principle. The change in momentum of Q is $m_2\mathbf{v}_2 - m_2\mathbf{u}_2$, with $m_2 = 4m$ and $u_2 = -2u$.

Substitute the value of v_2 from part **a**.

However the impulse is given as $\frac{80mu}{9}$.

So $\frac{60mu}{7}(1 + e) = \frac{80mu}{9}$.

$$(1 + e) = \frac{28}{27}$$

$$e = \frac{1}{27}$$

Form an equation and rearrange to find the value of e.

Exercise 4B

1 In each part of this question the two diagrams show the speeds of two particles A and B just before and just after a collision. The particles move on a smooth horizontal plane. Find the coefficient of restitution e in each case.

	Before collision	After collision
a	$6\,\mathrm{m\,s^{-1}}$ → (A) At rest (B)	At rest (A) $4\,\mathrm{m\,s^{-1}}$ → (B)
b	$4\,\mathrm{m\,s^{-1}}$ → (A) $2\,\mathrm{m\,s^{-1}}$ → (B)	$2\,\mathrm{m\,s^{-1}}$ → (A) $3\,\mathrm{m\,s^{-1}}$ → (B)
c	$9\,\mathrm{m\,s^{-1}}$ → (A) $6\,\mathrm{m\,s^{-1}}$ ← (B)	$3\,\mathrm{m\,s^{-1}}$ ← (A) $2\,\mathrm{m\,s^{-1}}$ → (B)

2 In each part of this question the two diagrams show the speeds of two particles A and B just before and just after a collision. The particles move on a smooth horizontal plane. The masses of A and B and the coefficients of restitution e are also given. Find the values of v_1 and v_2 in each case.

	Before collision	After collision
a $e=\frac{1}{2}$	$6\,\mathrm{m\,s^{-1}}$ → $A\,(0.25\,\mathrm{kg})$ At rest $B\,(0.5\,\mathrm{kg})$	$v_1\,\mathrm{m\,s^{-1}}$ → $A\,(0.25\,\mathrm{kg})$ $v_2\,\mathrm{m\,s^{-1}}$ → $B\,(0.5\,\mathrm{kg})$
b $e=0.25$	$4\,\mathrm{m\,s^{-1}}$ → $A\,(2\,\mathrm{kg})$ $2\,\mathrm{m\,s^{-1}}$ → $B\,(3\,\mathrm{kg})$	$v_1\,\mathrm{m\,s^{-1}}$ → $A\,(2\,\mathrm{kg})$ $v_2\,\mathrm{m\,s^{-1}}$ → $B\,(3\,\mathrm{kg})$
c $e=\frac{1}{7}$	$8\,\mathrm{m\,s^{-1}}$ → $A\,(3\,\mathrm{kg})$ $6\,\mathrm{m\,s^{-1}}$ ← $B\,(1\,\mathrm{kg})$	$v_1\,\mathrm{m\,s^{-1}}$ → $A\,(3\,\mathrm{kg})$ $v_2\,\mathrm{m\,s^{-1}}$ → $B\,(1\,\mathrm{kg})$
d $e=\frac{2}{3}$	$6\,\mathrm{m\,s^{-1}}$ → $A\,(400\,\mathrm{g})$ $6\,\mathrm{m\,s^{-1}}$ ← $B\,(400\,\mathrm{g})$	$v_1\,\mathrm{m\,s^{-1}}$ → $A\,(400\,\mathrm{g})$ $v_2\,\mathrm{m\,s^{-1}}$ → $B\,(400\,\mathrm{g})$
e $e=\frac{1}{5}$	$3\,\mathrm{m\,s^{-1}}$ → $A\,(5\,\mathrm{kg})$ $12\,\mathrm{m\,s^{-1}}$ ← $B\,(4\,\mathrm{kg})$	$v_1\,\mathrm{m\,s^{-1}}$ → $A\,(5\,\mathrm{kg})$ $v_2\,\mathrm{m\,s^{-1}}$ → $B\,(4\,\mathrm{kg})$

3 A small smooth sphere A of mass 1 kg is travelling along a straight line on a smooth horizontal plane with speed 4 m s^{-1} when it collides with a second smooth sphere B of the same radius, with mass 2 kg and travelling in the same direction as A with speed 2.5 m s^{-1}. After the collision, A continues in the same direction with speed 2 m s^{-1}.

  a Find the speed of B after the collision.

  b Find the coefficient of restitution for the spheres.

4 Two spheres A and B are of equal radius and have masses 2 kg and 6 kg respectively. A and B move towards each other along the same straight line on a smooth horizontal surface with velocities 4 m s^{-1} and 6 m s^{-1} respectively. If the coefficient of ~~friction~~ restitution is $\frac{1}{5}$, find the velocities of the spheres after the collision and the magnitude of the impulse given to each sphere.

5 Two particles of masses $2m$ and $3m$ respectively are moving towards each other with speed u. If the $3m$ mass is brought to rest by the collision, find the speed of the $2m$ mass after the collision and the coefficient of restitution between the particles.

6 Two particles A and B are travelling along the same straight line in the same direction on a smooth horizontal surface with speeds $3u$ and u respectively. Particle A catches up and collides with particle B. If the mass of B is twice that of A and the coefficient of restitution is e find, in terms of e and u, expressions for the speeds of A and B after the collision.

7 Two identical particles of mass m are projected towards each other along the same straight line on a smooth horizontal surface with speeds of $2u$ and $3u$. After the collision the directions of motion of both particles are reversed. Show that this implies that the coefficient of restitution e satisfies the inequality $e > \frac{1}{5}$.

8 Two particles A and B of mass m and km respectively are placed on a smooth horizontal plane. Particle A is made to move on the plane with speed u so as to collide directly with B which is at rest. After the collision B moves with speed $\frac{3}{10}u$.

  a Find, in terms of u and the constant k, the speed of A after the collision.

  b By using Newton's Law of Restitution show that $\frac{7}{3} \leqslant k \leqslant \frac{17}{3}$.

9 Two particles A and B of mass m and $3m$ respectively are placed on a smooth horizontal plane. Particle A is made to move on the plane with speed $2u$ so as to collide directly with B which is moving in the same direction with speed u. After the collision B moves with speed ku, where k is a positive constant.

  a Find, in terms of u and the constant k, the speed of A after the collision.

  b By using Newton's Law of Restitution show that $\frac{5}{4} \leqslant k \leqslant \frac{3}{2}$.

10 A particle P of mass m is moving with speed $4u$ on a smooth horizontal plane. The particle collides directly with a particle Q of mass $3m$ moving with speed $2u$ in the same direction as P. The coefficient of restitution between P and Q is e.

  a Show that the speed of Q after the collision is $\frac{u}{2}(5 + e)$.

  b Find the speed of P after the collision, giving your answer in terms of e.

  c Show that the direction of motion of P is unchanged by the collision.

  d Given that the magnitude of the impulse of P on Q is $2mu$, find the value of e.

4.3 You can also apply Newton's Law of Restitution to problems involving direct collision of a particle with a smooth plane surface perpendicular to the direction of motion of the particle.

In the figure a particle is shown moving horizontally with speed u before impact with a vertical plane surface. After impact the particle moves in the opposite direction with speed v.

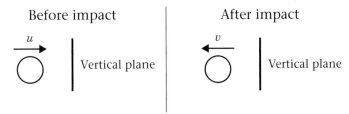

The speed of the particle after the impact depends on the speed of the particle before the impact and the coefficient of restitution, e, between the particle and the plane.

For the direct collision of a particle with a smooth plane Newton's Law of Restitution can be written as:

$$e = \frac{\text{speed of separation}}{\text{speed of approach}}$$

As the plane does not move:

$$e = \frac{v - 0}{u - 0} = \frac{v}{u}$$

■ $e = \dfrac{\text{speed of rebound}}{\text{speed of approach}}$

■ $e = \dfrac{v}{u}$

> You can also use
> $v = eu$ or $u = \dfrac{v}{e}$

Example 9

A particle collides normally with a fixed vertical plane. The diagram shows the speeds of the particle before and after the collision. Find the value of the coefficient of restitution e.

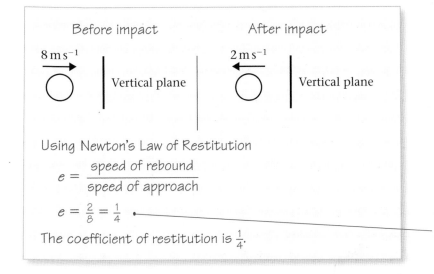

Using Newton's Law of Restitution

$$e = \frac{\text{speed of rebound}}{\text{speed of approach}}$$

$$e = \frac{2}{8} = \frac{1}{4}$$

The coefficient of restitution is $\frac{1}{4}$.

This is the same as
$$e = \frac{2 - 0}{8 - 0}.$$

Example 10

A small sphere collides normally with a fixed vertical wall. Before the impact the sphere is moving with a speed of $4\,\mathrm{m\,s^{-1}}$ on a smooth horizontal floor. The coefficient of restitution between the sphere and the wall is 0.2. Find the speed of the sphere after the collision.

Using Newton's Law of Restitution

$$e = \frac{\text{speed of rebound}}{\text{speed of approach}}$$

$$0.2 = \frac{v}{4}$$ — Let the speed of the sphere after the collision be $v\,\mathrm{m\,s^{-1}}$.

So $v = 0.8$ — Make v the subject of the formula.

The speed of the sphere after the collision is $0.8\,\mathrm{m\,s^{-1}}$

Example 11

A particle falls 22.5 cm from rest onto a smooth horizontal plane. It then rebounds to a height of 10 cm. Find the coefficient of restitution between the particle and the plane. Give your answer to 2 s.f.

<u>As particle falls:</u>

Use $v^2 = u^2 + 2as$ — The particle is falling under gravity so use constant acceleration formulae to find its speed when it hits the plane.

with $u = 0$, $s = 0.225$ and $a = g$

$v^2 = 0.45g$

$v = 2.1$ — $v\,\mathrm{m\,s^{-1}}$ is the speed of the particle when it reaches the plane.

<u>After impact:</u>

Use $v^2 = u^2 + 2as$

with $v = 0$, $s = 0.1$ and $a = -g$ — After it rebounds it initially moves upwards under gravity. As the upward direction is taken as positive here, the acceleration is negative.

$u^2 = 0.2g$

$u = 1.4$ — $u\,\mathrm{m\,s^{-1}}$ is the rebound speed of the particle.

Using Newton's Law of Restitution

$$e = \frac{\text{speed of rebound}}{\text{speed of approach}}$$ — e is the coefficient of restitution.

$$= \frac{1.4}{2.1} = \frac{2}{3}$$

The coefficient of restitution is 0.67 (2 s.f.)

Exercise 4C

Whenever a numerical value of g is required, take $g = 9.8 \, \text{m s}^{-2}$.

1 A smooth sphere collides normally with a fixed vertical wall. The two diagrams show the speeds of the sphere before and after collision. In each case find the value of the coefficient of restitution e.

a

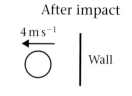

b

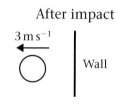

2 A smooth sphere collides normally with a fixed vertical wall. The two diagrams show the speed of the sphere before and after the collision. The value of e is given in each case. Find the speed of the sphere after the collision in each case.

a $e = \frac{1}{2}$

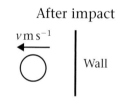

b $e = \frac{1}{4}$

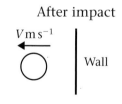

3 A smooth sphere collides normally with a fixed vertical wall. The two diagrams show the speed of the sphere before and after the collision. The value of e is given in each case. Find the speed of the sphere before the collision in each case.

a $e = \frac{1}{2}$

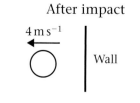

b $e = \frac{3}{4}$

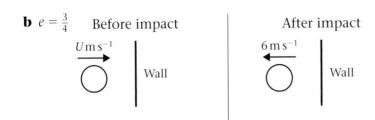

Before impact

After impact

$U\,\mathrm{m\,s}^{-1}$

Wall

$6\,\mathrm{m\,s}^{-1}$

Wall

4 A small smooth sphere of mass 0.3 kg is moving on a smooth horizontal table with a speed of $10\,\mathrm{m\,s}^{-1}$ when it collides normally with a fixed smooth wall. It rebounds with a speed of $7.5\,\mathrm{m\,s}^{-1}$. Find the coefficient of restitution between the sphere and the wall.

5 A particle falls 2.5 m from rest on to a smooth horizontal plane. It then rebounds to a height of 1.5 m. Find the coefficient of restitution between the particle and the plane. Give your answer to 2 significant figures.

6 A particle falls 3 m from rest onto a smooth horizontal plane. It then rebounds to a height h m. The coefficient of restitution between the particle and the plane is 0.25. Find the value of h.

7 A small smooth sphere falls from rest onto a smooth horizontal plane. It takes 2 seconds to reach the plane then another 2 seconds to reach the plane a second time. Find the coefficient of restitution between the particle and the plane.

8 A small smooth sphere falls from rest onto a smooth horizontal plane. It takes 3 seconds to reach the plane. The coefficient of restitution between the particle and the plane is 0.49. Find the time it takes for the sphere to reach the plane a second time.

4.4 You can solve problems relating to successive impacts involving three particles, or two particles and a smooth plane surface by considering each collision separately. You can also solve problems relating to successive bounces on a horizontal plane.

When you are solving problems involving successive impacts, you should draw a clear diagram showing the 'before' and 'after' information for each collision.

Example 12

Three spheres A, B and C have masses m, $2m$ and $3m$ respectively. The spheres move along the same straight line on a horizontal plane with A following B, which is following C. Initially the velocities of A, B and C are $7\,\mathrm{m\,s}^{-1}$, $3\,\mathrm{m\,s}^{-1}$ and $1\,\mathrm{m\,s}^{-1}$ respectively, in the direction ABC. Sphere A collides with sphere B and then sphere B collides with sphere C. The coefficient of restitution between A and B is $\frac{1}{2}$ and between B and C is $\frac{1}{4}$.

a Find the velocities of the three spheres after the second collision.

b Explain how you can predict that there will be a further collision between A and B.

a <u>First collision</u>

Before impact between A and B

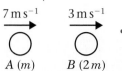

$7\,\text{m}\,\text{s}^{-1}$ $3\,\text{m}\,\text{s}^{-1}$

A (m)   B (2m)

> Draw diagrams to show the masses and velocities of A and B before and after the first collision.

After impact between A and B

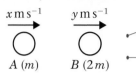

$x\,\text{m}\,\text{s}^{-1}$ $y\,\text{m}\,\text{s}^{-1}$

A (m)   B (2m)

> Let the velocity of A be $x\,\text{m}\,\text{s}^{-1}$ and let the velocity of B be $y\,\text{m}\,\text{s}^{-1}$ after the impact. Make it clear in your diagram which velocity corresponds to which particle.

Using conservation of momentum ($\rightarrow$)

$$m \times 7 + 2m \times 3 = m \times x + 2m \times y$$
$$\Rightarrow x + 2y = 13 \qquad ❶$$

> Use
> $m_1\mathbf{u}_1 + m_2\mathbf{u}_2 = m_1\mathbf{v}_1 + m_2\mathbf{v}_2.$

Using Newton's Law of Restitution

$$\frac{y - x}{7 - 3} = \frac{1}{2}$$
$$\Rightarrow y - x = 2 \qquad ❷$$

> Use $\dfrac{\text{speed of separation of particles}}{\text{speed of approach of particles}} = e$

Solving equations ❶ and ❷ gives

$$y = 5 \text{ and } x = 3.$$

> Add equations ❶ and ❷ to eliminate x and to give $3y = 15$, then substitute $y = 5$ into equation ❷ to give $x = 3$.

After the first impact the speed of A is $3\,\text{m}\,\text{s}^{-1}$ and the speed of B is $5\,\text{m}\,\text{s}^{-1}$

<u>Second collision</u>

Before impact between B and C

$5\,\text{m}\,\text{s}^{-1}$ $1\,\text{m}\,\text{s}^{-1}$

B (2m)   C (3m)

> Draw diagrams to show the masses and velocities of B and C before and after the second collision.

After impact between B and C

$v\,\text{m}\,\text{s}^{-1}$ $w\,\text{m}\,\text{s}^{-1}$

B (2m)   C (3m)

> Let the velocity of B be $v\,\text{m}\,\text{s}^{-1}$ and let the velocity of C be $w\,\text{m}\,\text{s}^{-1}$ after the impact. Use different letters for the speeds after this collision (i.e. not x and y again).

Using conservation of momentum ($\rightarrow$)

$$2m \times 5 + 3m \times 1 = 2m \times v + 3m \times w$$
$$\Rightarrow 2v + 3w = 13 \qquad ❸$$

Using Newton's Law of Restitution

$$\frac{w - v}{5 - 1} = \frac{1}{4}$$
$$\Rightarrow w - v = 1 \qquad ❹$$

> Form two equations and solve them as you did for the first collision.

Solving equations ❸ and ❹ gives

$$w = 3 \text{ and } v = 2.$$

After the second impact the velocity of B is
$2\,\text{m s}^{-1}$ and the velocity of C is $3\,\text{m s}^{-1}$.
The velocity of A is $3\,\text{m s}^{-1}$.
These are all in the direction ABC.

The velocity of A became $3\,\text{m s}^{-1}$ as a result of the first collision. Its velocity is unchanged by the second collision.

b As the velocity of A is greater than, and in the same direction as the velocity of B there will be a further collision between A and B.

Example 13

A uniform smooth sphere P of mass $3m$ is moving in a straight line with speed u on a smooth horizontal table. Another uniform smooth sphere Q, of mass m and having the same radius as P, is moving with speed $2u$ in the same straight line as P, but in the opposite direction. The sphere P collides with the sphere Q directly. The velocities of P and Q after the collision are v and w respectively, measured in the direction of motion of P before the collision. The coefficient of restitution between P and Q is e.

a Find expressions for v and w in terms of u and e.

b Show that, if the direction of motion of P is changed by the collision, then $e > \frac{1}{3}$.

Following the collision with P, the sphere Q then collides with and rebounds from a vertical wall, which is perpendicular to the direction of motion of Q. The coefficient of restitution between Q and the wall is e'.

c Given that $e = \frac{5}{9}$ and that P and Q collide again in the subsequent motion show that $e' > \frac{1}{9}$. **E**

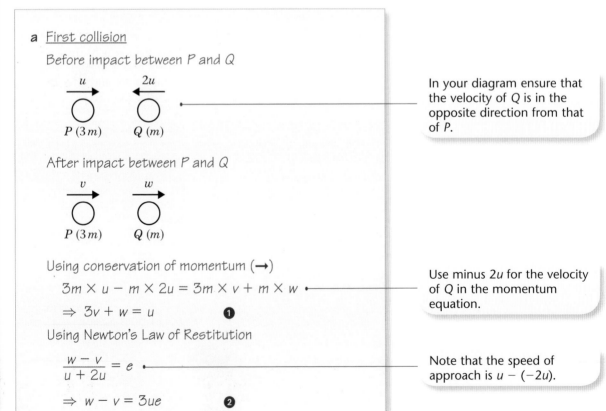

a First collision

Before impact between P and Q

u $2u$

$P\,(3m)$ $Q\,(m)$

In your diagram ensure that the velocity of Q is in the opposite direction from that of P.

After impact between P and Q

v w

$P\,(3m)$ $Q\,(m)$

Using conservation of momentum ($\rightarrow$)

$3m \times u - m \times 2u = 3m \times v + m \times w$

$\Rightarrow 3v + w = u$　　**1**

Use minus $2u$ for the velocity of Q in the momentum equation.

Using Newton's Law of Restitution

$\dfrac{w - v}{u + 2u} = e$

$\Rightarrow w - v = 3ue$　　**2**

Note that the speed of approach is $u - (-2u)$.

Subtract equation ❷ from equation ❶ to give

$$4v = u(1 - 3e)$$

$$v = \frac{u(1 - 3e)}{4}$$

$$w = \frac{u(9e + 1)}{4}$$

Solve the two simultaneous equations to obtain expressions for v and w.

Substitute v into equation ❷ to find the expression for w.

b As $v < 0$, $\dfrac{u(1 - 3e)}{4} < 0$

So $e > \frac{1}{3}$

As P changes direction it moves to the left and $v < 0$.

c Before impact between Q and the wall

w

Q (m)

After impact between Q and the wall

V

Q (m)

Draw a diagram for the new collision.

Using Newton's Law of Restitution

$$\frac{V}{w} = e'$$

Use $\dfrac{\text{speed of rebound}}{\text{speed of approach}} = e$

with $e = e'$.

Using $e = \dfrac{5}{9}$, $w = \dfrac{u(5 + 1)}{4} = \dfrac{3u}{2}$

So $V = \dfrac{3ue'}{2}$

Use your answer to part **a** to find w and use Newton's Law of Restitution to find V.

Also when $e = \dfrac{5}{9}$, $v = -\dfrac{u}{6}$

Also evaluate v using your answer to part **a** and $e = \frac{5}{9}$.

After impact between Q and wall

$\frac{u}{6}$ $\frac{3ue'}{2}$

P $(3m)$ Q (m)

Another diagram is helpful here.

Since Q and P collide again

$$\frac{3ue'}{2} > \frac{u}{6}$$

So $e' > \frac{1}{9}$.

Use the condition that speed of Q > speed of P for collision.

Example 14

A tennis ball, which may be modelled as a particle, is dropped from rest at a height of 90 cm onto a smooth horizontal plane. The coefficient of restitution between the ball and the plane is 0.5. Assume that there is no air resistance and that the ball falls under gravity and hits the plane at right angles.

a Find the height to which the ball rebounds after the first bounce.

b Find the height to which the ball rebounds after the second bounce.

c Find the total distance travelled by the ball before it comes to rest.

a As the tennis ball falls:

Use $v^2 = u^2 + 2as$

with $u = 0$, $s = 0.9$ and $a = g$

$$v^2 = 1.8g$$

$$v = 4.2$$

> The tennis ball is falling under gravity so use the appropriate constant acceleration formula to find the ball's speed when it hits the plane.

> $v\,\mathrm{m\,s^{-1}}$ is the speed of the particle when it hits the plane.

After first impact with plane:

By Newton's Law of Restitution

$$e = \frac{\text{speed of rebound}}{\text{speed of approach}}$$

> e is the coefficient of restitution.

$$0.5 = \frac{v'}{4.2}$$

$$v' = 2.1$$

> $v'\,\mathrm{m\,s^{-1}}$ is the rebound speed of the particle.

As the ball moves under gravity after impact:

Use $v^2 = u^2 + 2as$

with $v = 0$, $u = 2.1$ and $a = -g$

$$0 = 2.1^2 - 2gh_1$$

$$h_1 = \frac{2.1^2}{2g}$$

$$= 0.225$$

> After it rebounds it moves up a distance under gravity. As the upward direction is taken as positive here, the acceleration is minus g. Let $s = h_1$ when $v = 0$.

The ball rebounds after the first bounce to a height of 22.5 cm

b As the tennis ball falls:

$$v = v' = 2.1$$

After second impact:

By Newton's Law of Restitution

$$v'' = e \times 2.1 = 1.05$$

As the ball moves under gravity after the second impact:

> From symmetry $v'\,\mathrm{m\,s^{-1}}$ is the speed of the particle on its approach to the plane the second time.

> Let v'' be the speed of the ball after the second bounce.

Use $v^2 = u^2 + 2as$

with $v = 0$, $u = 1.05$ and $a = -g$

$$0 = 1.05^2 - 2gh_2$$

$$h_2 = \frac{1.05^2}{2g} = 0.05625$$

The ball rebounds after the second bounce to a height of 5.625 cm.

c The total distance travelled is

$$0.9 + 0.225 + 0.225 + 0.05625 + 0.05625 + \ldots$$

Distance travelled =

$$0.9 + 2(0.225 + 0.225 \times \tfrac{1}{4} + 0.225 \times \left(\tfrac{1}{4}\right)^2$$
$$+ 0.225 \times \left(\tfrac{1}{4}\right)^3 + \ldots)$$

$$= 0.9 + 2 \times \frac{0.225}{\left(1 - \tfrac{1}{4}\right)} = 1.5$$

The total distance travelled by the ball before it comes to rest is 1.5 m.

After it rebounds it moves under gravity to a height h_2. Again the acceleration is minus g. Let $v = 0$ when $s = h_2$.

If you repeat this working you find that on subsequent bounces:
$$h_3 = \tfrac{1}{4}h_2, \; h_4 = \tfrac{1}{4}h_3, \ldots$$

The ball moves 0.9 m before first impact, then moves up 0.225 m and down 0.225 m before second impact then moves up 0.05625 m and down 0.05625 m before third impact.

The terms in the bracket form an infinite geometric progression.

Use the formula for the sum of an infinite G.P.
$$S = \frac{a}{1 - r}, \text{ with } a = 0.225$$
and $r = \tfrac{1}{4}$.

Exercise 4D

Whenever a numerical value of g is required, take $g = 9.8 \, \mathrm{m\,s^{-2}}$.

1 Three small smooth spheres A, B and C of equal radius move along the same straight line on a horizontal plane. Sphere A collides with sphere B and then sphere B collides with sphere C. The diagrams show the velocities before the first collision, after the first collision between A and B and then after the collision between B and C.

a Find the values of u, v, x and y if $e = \tfrac{1}{2}$ for both collisions.

Before collision			After A and B have collided			After B and C have collided		
$5\,\mathrm{m\,s^{-1}}$	$1\,\mathrm{m\,s^{-1}}$	$4\,\mathrm{m\,s^{-1}}$	$u\,\mathrm{m\,s^{-1}}$	$v\,\mathrm{m\,s^{-1}}$	$4\,\mathrm{m\,s^{-1}}$	$u\,\mathrm{m\,s^{-1}}$	$x\,\mathrm{m\,s^{-1}}$	$y\,\mathrm{m\,s^{-1}}$
A (2 kg)	B (1 kg)	C (2 kg)	A (2 kg)	B (1 kg)	C (2 kg)	A (2 kg)	B (1 kg)	C (2 kg)

b Find the values of u, v, x and y if $e = \tfrac{1}{6}$ for the collision between A and B and $e = \tfrac{1}{2}$ for the collision between B and C.

Before collision			After A and B have collided			After B and C have collided		
$10\,\mathrm{m\,s^{-1}}$	$2\,\mathrm{m\,s^{-1}}$	$3\,\mathrm{m\,s^{-1}}$	$u\,\mathrm{m\,s^{-1}}$	$v\,\mathrm{m\,s^{-1}}$	$3\,\mathrm{m\,s^{-1}}$	$u\,\mathrm{m\,s^{-1}}$	$x\,\mathrm{m\,s^{-1}}$	$y\,\mathrm{m\,s^{-1}}$
A (1.5 kg)	B (2 kg)	C (1 kg)	A (1.5 kg)	B (2 kg)	C (1 kg)	A (1.5 kg)	B (2 kg)	C (1 kg)

2 Three small smooth spheres A, B and C of equal radius have masses 500 g, 500 g and 1 kg respectively. The spheres move along the same straight line on a horizontal plane with A following B which is following C. Initially the velocities of A, B and C are $4\mathbf{i}$ m s^{-1}, $-2\mathbf{i}$ m s^{-1} and $0.5\mathbf{i}$ m s^{-1} respectively, where $\mathbf{i}$ is a unit vector in the direction ABC. Sphere A collides with sphere B and then sphere B collides with sphere C. The coefficient of restitution between A and B is $\frac{2}{3}$ and between B and C is $\frac{1}{2}$. Find the velocities of the three spheres after all of the collisions have taken place.

3 Three perfectly elastic particles A, B and C of masses $3m$, $5m$ and $4m$ respectively lie at rest on a straight line on a smooth horizontal table with B between A and C. Particle A is projected directly towards B with speed 6 m s^{-1} and after A has collided with B, B then collides with C. Find the speed of each particle after the second impact.

4 Three identical smooth spheres A, B and C, each of mass m, lie at rest on a straight line on a smooth horizontal table. Sphere A is projected with speed u to strike sphere B directly. Sphere B then strikes sphere C directly. The coefficient of restitution between any two spheres is e, $e \neq 1$.

a Find the speeds in terms of u and e of the spheres after these two collisions.

b Show that A will catch up with B and there will be a further collision.

5 Three identical spheres A, B and C of equal mass m, and equal radius move along the same straight line on a horizontal plane. B is between A and C. A and B are moving towards each other with velocities $4u$ and $2u$ respectively while C moves away from B with velocity $3u$.

a If the coefficient of restitution between any two of the spheres is e, show that B will only collide with C if $e > \frac{2}{3}$.

b Find the direction of motion of A after collision, if $e > \frac{2}{3}$.

6 Two particles P of mass $2m$ and Q of mass $3m$ are moving towards each other with speeds $4u$ and $2u$ respectively. The direction of motion of Q is reversed by the impact and its speed after impact is u. This particle then hits a smooth vertical wall perpendicular to its direction of motion. The coefficient of restitution between Q and the wall is $\frac{2}{3}$. In the subsequent motion, there is a further collision between Q and P. Find the speeds of P and Q after this collision.

7 Two small smooth spheres P and Q of equal radius have masses m and $3m$ respectively. Sphere P is moving with speed $12u$ on a smooth horizontal table when it collides directly with Q which is at rest on the table. The coefficient of restitution between P and Q is $\frac{2}{3}$.

a Find the speeds of P and Q immediately after the collision.

After the collision Q hits a smooth vertical wall perpendicular to the direction of its motion. The coefficient of restitution between Q and the wall is $\frac{4}{5}$.
Q then collides with P a second time.

b Find the speeds of P and Q after the second collision between P and Q.

8 A small smooth table tennis ball, which may be modelled as a particle, falls from rest at a height 40 cm onto a smooth horizontal plane. The coefficient of restitution between the ball and the plane is 0.7.

a Find the height to which the ball rebounds after the first bounce.

b Find the height to which the ball rebounds after the second bounce.

c Find the total distance travelled by the ball before it comes to rest.

9 A small smooth ball, which may be modelled as a particle, falls from rest at a height H onto a smooth horizontal plane. The coefficient of restitution between the ball and the plane is e.

a Find in terms of H and e the height to which the ball rebounds after the first bounce.

b Find in terms of H and e the height to which the ball rebounds after the second bounce.

c Find an expression for the total distance travelled by the ball before it comes to rest.

4.5 You can solve problems which ask you to find the change in energy due to an impact or the application of an impulse.

Example 15

Two spheres A and B are of equal radii and have masses 3 kg and 5 kg respectively. A and B move towards each other along the same straight line on a smooth horizontal surface with velocities $3\,\text{m s}^{-1}$ and $2\,\text{m s}^{-1}$ respectively.

a If the coefficient of restitution is $\frac{3}{5}$ find the velocities of the spheres after the collision.

b Find also the loss of kinetic energy due to the impact.

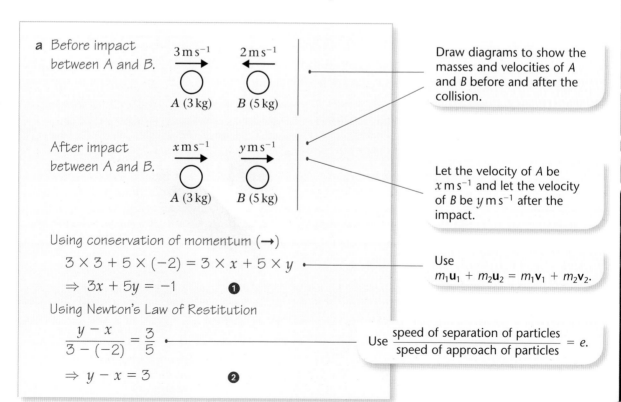

a Before impact between A and B.

$3\,\text{m s}^{-1}$ $2\,\text{m s}^{-1}$

A (3 kg)   B (5 kg)

Draw diagrams to show the masses and velocities of A and B before and after the collision.

After impact between A and B.

$x\,\text{m s}^{-1}$ $y\,\text{m s}^{-1}$

A (3 kg)   B (5 kg)

Let the velocity of A be $x\,\text{m s}^{-1}$ and let the velocity of B be $y\,\text{m s}^{-1}$ after the impact.

Using conservation of momentum ($\rightarrow$)

$3 \times 3 + 5 \times (-2) = 3 \times x + 5 \times y$

$\Rightarrow 3x + 5y = -1$ **❶**

Use $m_1u_1 + m_2u_2 = m_1v_1 + m_2v_2$.

Using Newton's Law of Restitution

$\dfrac{y - x}{3 - (-2)} = \dfrac{3}{5}$

$\Rightarrow y - x = 3$ **❷**

Use $\dfrac{\text{speed of separation of particles}}{\text{speed of approach of particles}} = e$.

Solving equations **1** and **2** gives

$$y = 1 \text{ and } x = -2.$$

After the impact the direction of A is reversed and its speed is $2\,\text{m s}^{-1}$. The direction of B is also reversed and its speed is $1\,\text{m s}^{-1}$.

b The total kinetic energy before impact is

$$\tfrac{1}{2} \times 3 \times 3^2 + \tfrac{1}{2} \times 5 \times 2^2 = 23.5\,\text{J}$$

The total kinetic energy after impact is

$$\tfrac{1}{2} \times 3 \times 2^2 + \tfrac{1}{2} \times 5 \times 1^2 = 8.5\,\text{J}$$

So the loss of kinetic energy is

$$23.5\,\text{J} - 8.5\,\text{J} = 15\,\text{J}.$$

> Solve the simultaneous equations **1** and **2** to find x and y.

> The total kinetic energy before impact is $\tfrac{1}{2}m_1u_1^2 + \tfrac{1}{2}m_2u_2^2$.

> The total kinetic energy after impact is $\tfrac{1}{2}m_1v_1^2 + \tfrac{1}{2}m_2v_2^2$.

■ **The loss of kinetic energy due to impact is**
$$(\tfrac{1}{2}m_1u_1^2 + \tfrac{1}{2}m_2u_2^2) - (\tfrac{1}{2}m_1v_1^2 + \tfrac{1}{2}m_2v_2^2)$$

> When the particles are perfectly elastic ($e = 1$) you will find that there is no loss of kinetic knergy due to impact.
> In all practical situations $e < 1$ and some kinetic energy is converted into heat or sound energy at impact.

Example 16

A gun of mass $600\,\text{kg}$ fires a shell of mass $12\,\text{kg}$ horizontally with velocity $20\,\text{m s}^{-1}$.

a Find the velocity of the gun after the shell has been fired.

b Find the total kinetic energy generated on firing.

c Show that the ratio of the energy of the gun to the energy of the shell is equal to the ratio of the speed of the gun to the speed of the shell after firing.

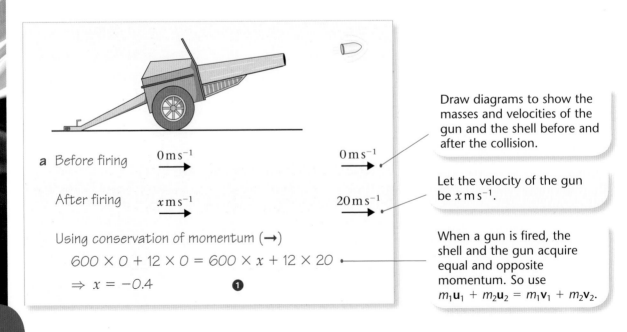

a Before firing $0\,\text{m s}^{-1}$ $0\,\text{m s}^{-1}$

    After firing $x\,\text{m s}^{-1}$ $20\,\text{m s}^{-1}$

Using conservation of momentum (→)

$$600 \times 0 + 12 \times 0 = 600 \times x + 12 \times 20$$
$$\Rightarrow x = -0.4 \qquad \mathbf{1}$$

> Draw diagrams to show the masses and velocities of the gun and the shell before and after the collision.

> Let the velocity of the gun be $x\,\text{m s}^{-1}$.

> When a gun is fired, the shell and the gun acquire equal and opposite momentum. So use $m_1u_1 + m_2u_2 = m_1v_1 + m_2v_2$.

After the impact the direction of the gun is reversed and its speed is $0.4\,\text{m s}^{-1}$.

b The total kinetic energy after firing is

$$\tfrac{1}{2} \times 600 \times 0.4^2 + \tfrac{1}{2} \times 12 \times 20^2$$
$$= 48 + 2400$$
$$= 2448\,\text{J}$$

So the total kinetic energy generated on firing is 2448 Joules.

Both the gun and the shell acquire kinetic energy supplied by expanding gas resulting from the chemical reaction that takes place on firing.

c The ratio   K.E. of gun: K.E. of shell = 1:50

The ratio   speed of recoil of gun:
             speed of projection of shell = 1:50

These ratios are equal.

Note that the shell has a much higher energy than the gun.

Example 17

A tennis ball of mass 0.2 kg is moving with velocity $(-12\mathbf{i} - 2\mathbf{j})\,\text{m s}^{-1}$ when it is struck by a tennis racquet. This is modelled as an impulse of $(5.6\mathbf{i} + 2.8\mathbf{j})\,\text{Ns}$. Find the resulting velocity of the tennis ball and the kinetic energy gained by the ball as a result of the impact.

The change in momentum of the particle is

$$0.2\mathbf{v} - 0.2(-12\mathbf{i} - 2\mathbf{j})\,\text{kg m s}^{-1}.$$

From the impulse–momentum principle this is equal to the impulse.

$$0.2\mathbf{v} - 0.2(-12\mathbf{i} - 2\mathbf{j}) = 5.6\mathbf{i} + 2.8\mathbf{j}$$
$$0.2\mathbf{v} = 5.6\mathbf{i} + 2.8\mathbf{j} - 2.4\mathbf{i} - 0.4\mathbf{j}$$
$$= 3.2\mathbf{i} + 2.4\mathbf{j}$$
$$\mathbf{v} = 16\mathbf{i} + 12\mathbf{j}$$

Change in kinetic energy =

$$\tfrac{1}{2} \times 0.2 \times (16^2 + 12^2) - \tfrac{1}{2} \times 0.2 \times ((-12)^2 + (-2)^2)$$
$$= 40 - 14.8 = 25.2\,\text{J}$$

So the resulting velocity is $(16\mathbf{i} + 12\mathbf{j})\,\text{m s}^{-1}$ and the gain in K.E. is 25.2 joules.

Let the velocity of the particle after the impact be $\mathbf{v}\,\text{m s}^{-1}$.

Use $m\mathbf{v} - m\mathbf{u} = \mathbf{I}$ substituting $m = 0.2$, $\mathbf{u} = -12\mathbf{i} - 2\mathbf{j}$ and $\mathbf{I} = 5.6\mathbf{i} + 2.8\mathbf{j}$.

Use kinetic energy $= \tfrac{1}{2} \times$ mass $\times$ speed^2.

Example 18

Two particles, A and B, of mass 200 g and 300 g respectively, are connected by a light inextensible string. The particles are side by side on a smooth floor and A is projected with speed $6\,\mathrm{m\,s^{-1}}$ directly away from B. When the string becomes taut particle B is jerked into motion and A and B then move with a common speed in the direction of the original velocity of A. Find

a the common speed of the particles after the string becomes taut,

b the loss of total kinetic energy due to the jerk.

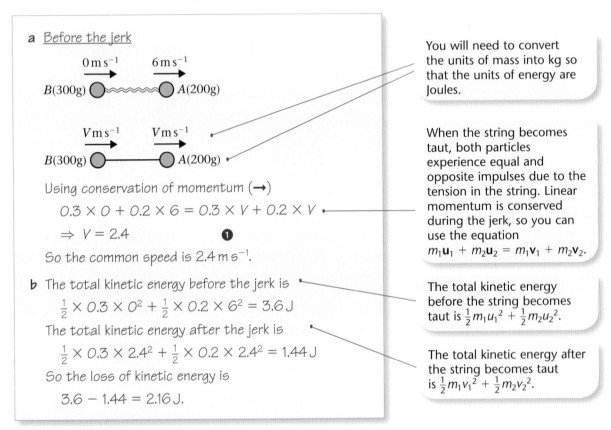

a Before the jerk

$0\,\mathrm{m\,s^{-1}}$ $6\,\mathrm{m\,s^{-1}}$

$B(300\mathrm{g})$ $A(200\mathrm{g})$

$V\,\mathrm{m\,s^{-1}}$ $V\,\mathrm{m\,s^{-1}}$

$B(300\mathrm{g})$ $A(200\mathrm{g})$

Using conservation of momentum ($\rightarrow$)

$0.3 \times 0 + 0.2 \times 6 = 0.3 \times V + 0.2 \times V$

$\Rightarrow V = 2.4$ ①

So the common speed is $2.4\,\mathrm{m\,s^{-1}}$.

b The total kinetic energy before the jerk is

$\frac{1}{2} \times 0.3 \times 0^2 + \frac{1}{2} \times 0.2 \times 6^2 = 3.6\,\mathrm{J}$

The total kinetic energy after the jerk is

$\frac{1}{2} \times 0.3 \times 2.4^2 + \frac{1}{2} \times 0.2 \times 2.4^2 = 1.44\,\mathrm{J}$

So the loss of kinetic energy is

$3.6 - 1.44 = 2.16\,\mathrm{J}$.

> You will need to convert the units of mass into kg so that the units of energy are Joules.

> When the string becomes taut, both particles experience equal and opposite impulses due to the tension in the string. Linear momentum is conserved during the jerk, so you can use the equation $m_1\mathbf{u}_1 + m_2\mathbf{u}_2 = m_1\mathbf{v}_1 + m_2\mathbf{v}_2$.

> The total kinetic energy before the string becomes taut is $\frac{1}{2}m_1u_1^2 + \frac{1}{2}m_2u_2^2$.

> The total kinetic energy after the string becomes taut is $\frac{1}{2}m_1v_1^2 + \frac{1}{2}m_2v_2^2$.

Exercise 4E

1 A particle A of mass 500 g lies at rest on a smooth horizontal table. A second particle B of mass 600 g is projected along the table with velocity $6\,\mathrm{m\,s^{-1}}$ and collides directly with A. If the collision reduces the speed of B to $1\,\mathrm{m\,s^{-1}}$, without changing its direction, find

a the speed of A after the collision,

b the loss of kinetic energy due to the collision.

2 Two particles A and B of mass m and $2m$ respectively move towards each other with speeds u and $2u$. If the coefficient of restitution between the spheres is $\frac{2}{3}$, find the speeds of A and of B after the collision. Find also, in terms of m and u, the loss of kinetic energy due to the collision.

3 A particle of mass 3 kg moving with velocity $6\,\mathrm{m\,s^{-1}}$ collides directly with a particle of mass 5 kg moving in the opposite direction with velocity $2\,\mathrm{m\,s^{-1}}$. The particles coalesce and move with velocity v after the collision. Find the loss of kinetic energy due to the impact.

4 A billiard ball of mass 200 g strikes a smooth cushion at right angles. Its velocity before the impact is $2.5\,\mathrm{m\,s^{-1}}$ and the coefficient of restitution is $\frac{4}{5}$. Find the loss in kinetic energy of the billiard ball due to the impact.

5 A bullet of mass 0.15 kg moving horizontally at $402\,\mathrm{m\,s^{-1}}$ embeds itself in a sandbag of mass 30 kg, which is suspended freely. Assuming that the sandbag is stationary before the impact, find

   a the common velocity of the bullet and the sandbag,

   b the loss of kinetic energy due to the impact.

6 A particle of mass 0.4 kg is moving with velocity $(\mathbf{i} - 4\mathbf{j})\,\mathrm{m\,s^{-1}}$ when it receives an impulse $(3\mathbf{i} + 2\mathbf{j})\,\mathrm{Ns}$. Find the new velocity of the particle and the change in kinetic energy of the particle as a result of the impulse.

7 A squash ball of mass 0.025 kg is moving with velocity $(22\mathbf{i} + 37\mathbf{j})\,\mathrm{m\,s^{-1}}$ when it hits a wall. It rebounds with velocity $(10\mathbf{i} - 11\mathbf{j})\,\mathrm{m\,s^{-1}}$. Find the change in kinetic energy of the squash ball.

8 A particle of mass 0.2 kg is moving with velocity $(5\mathbf{i} + 25\mathbf{j})\,\mathrm{m\,s^{-1}}$ when it collides with a particle of mass 0.1 kg moving with velocity $(2\mathbf{i} + 10\mathbf{j})\,\mathrm{m\,s^{-1}}$. The two particles coalesce and form one particle of mass 0.3 kg. Find the velocity of the combined particle and find the loss in kinetic energy as a result of the collision.

9 A bullet is fired horizontally from a rifle. The rifle has mass 4.8 kg and the bullet has mass 20 g. The initial speed of the bullet is $400\,\mathrm{m\,s^{-1}}$. Find

   a the initial speed with which the rifle recoils,

   b the total kinetic energy generated as a result of firing the bullet.

10 A train of mass 30 tonnes moving with a small velocity V impacts upon a number of stationary carriages each weighing 6 tonnes. The complete train and carriages now move forward with a velocity of $\frac{5}{8}V$. Find

   a the number of stationary carriages,

   b the fraction of the original kinetic energy lost in the impact.

11 A truck of mass 5 tonnes is moving at $1.5\,\mathrm{m\,s^{-1}}$ when it hits a second truck of mass 10 tonnes which is at rest. After the impact the second truck moves at $0.6\,\mathrm{m\,s^{-1}}$. Find the speed of the first truck after the impact and the total loss of kinetic energy due to the impact.

12 A particle of mass m moves in a straight line with velocity v when it explodes into two parts, one of mass $\frac{1}{3}m$ and the other of mass $\frac{2}{3}m$ both moving in the same direction as before. If the explosion increases the energy of the system by $\frac{1}{4}mu^2$, where u is a positive constant, find the velocities of the particles immediately after the explosion. Give your answers in terms of u and v.

13 A small smooth sphere A of mass 2 kg moves at 4 m s^{-1} on a smooth horizontal table. It collides directly with a second equal-sized smooth sphere B of mass 3 kg, which is moving away from A in the same direction at a speed of 1 m s^{-1}. If the loss of kinetic energy due to the collision is 3 J find the speeds and the directions of the two spheres after the collision.

14 Two particles, A and B, of masses 2 kg and 5 kg respectively, are connected by a light inextensible string. The particles are side by side on a smooth floor and A is projected with speed 7 m s^{-1} directly away from B. When the string becomes taut particle B is jerked into motion and A and B then move with a common speed in the direction of the original velocity of A. Find

  a the common speed of the particles after the string becomes taut,

  b the loss of total kinetic energy due to the jerk.

15 Two particles, A and B, of masses m and M respectively, are connected by a light inextensible string. The particles are side by side on a smooth floor and A is projected with speed u directly away from B. When the string becomes taut particle B is jerked into motion and A and B then move with a common speed in the direction of the original velocity of A. Find the common speed of the particles after the string becomes taut, and show that the loss of total kinetic energy due to the jerk is $\dfrac{mMu^2}{2(m + M)}$.

16 Two particles of masses 3 kg and 5 kg lie on a smooth table and are connected by a slack inextensible string. The first particle is projected along the table with a velocity of 20 m s^{-1} directly away from the second particle.

  a Find the velocity of each particle after the string has become taut.

  b Find the difference between the kinetic energies of the system when the string is slack and when it is taut.

The second particle is attached to a third particle of unknown mass by another slack string, and the velocity of the whole system after both strings have become taut is 6 m s^{-1}.

  c Find the mass of the third particle.

17 Three small spheres of mass 20 g, 40 g and 60 g respectively lie in order in a straight line on a large smooth table. The distance between adjacent spheres is 10 cm. Two slack strings, each 70 cm in length, connect the first sphere with the second, and the second sphere with the third. The 60 g sphere is projected with a speed of 5 m s^{-1}, directly away from the other two.

  a Find the time which elapses before the 20 g sphere begins to move and the speed with which it starts.

  b Find the loss in kinetic energy resulting from the two jerks.

Mixed exercise 4F

1 A cricket ball of mass 0.5 kg is struck by a bat. Immediately before being struck the velocity of the ball is $-25\mathbf{i}$ m s^{-1}. Immediately after being struck the velocity of the ball is $(23\mathbf{i} + 20\mathbf{j})$ m s^{-1}. Find the magnitude of the impulse exerted on the ball by the bat and the angle between the impulse and the direction of $\mathbf{i}$.

2 A ball of mass 0.2 kg is hit by a bat which gives it an impulse of $(2.4\mathbf{i} + 3.6\mathbf{j})$ Ns. The velocity of the ball immediately after being hit is $(12\mathbf{i} + 5\mathbf{j})$ m s^{-1}. Find the velocity of the ball immediately before it is hit.

3 A particle P of mass 0.3 kg is moving so that its position vector $\mathbf{r}$ metres at time t seconds is given by
$$\mathbf{r} = (t^3 + t^2 + 4t)\mathbf{i} + (11t)\mathbf{j}$$
a Calculate the speed of P when $t = 4$.

When $t = 4$, the particle is given an impulse $(2.4\mathbf{i} + 3.6\mathbf{j})$ Ns.

b Find the velocity of P immediately after the impulse.

4 Two identical spheres, moving in opposite directions, collide directly. As a result of the impact one of the spheres is brought to rest. The coefficient of restitution between the spheres is $\frac{1}{3}$. Show that the ratio of the speeds of the spheres before the impact is $2:1$.

5 A particle P of mass m is moving in a straight line with speed $\frac{1}{4}u$ at the instant when it collides directly with a particle Q, of mass λm, which is at rest. The coefficient of restitution between P and Q is $\frac{1}{4}$. Given that P comes to rest immediately after hitting Q find the value of λ.

6 A boy of mass m dives off a boat of mass M which was previously at rest. Immediately after diving off, the boy has a horizontal speed of V. Calculate the speed with which the boat begins to move. Prove that the total kinetic energy of the boy and the boat is $\dfrac{m(m + M)V^2}{2M}$.

7 Two spheres P and Q of equal radius and masses 4 kg and 2 kg respectively are travelling towards each other along a straight line on a smooth horizontal surface. Initially, P has a speed of 5 m s^{-1} and Q has a speed of 3 m s^{-1}. After the collision the direction of Q is reversed and it is travelling at a speed of 2 m s^{-1}. Find the speed of P after the collision and the loss of kinetic energy due to the collision.

8 A body P of mass 4 kg is moving with velocity $(2\mathbf{i} + 16\mathbf{j})$ m s^{-1} when it collides with a body Q of mass 3 kg moving with velocity $(-\mathbf{i} - 8\mathbf{j})$ m s^{-1}. Immediately after the collision the velocity of P is $(-4\mathbf{i} - 32\mathbf{j})$ m s^{-1}. Find the velocity of Q immediately after the collision.

9 A particle P of mass $3m$ is moving in a straight line with speed u at the instant when it collides directly with a particle Q of mass m which is at rest. The coefficient of restitution between P and Q is e.

    a Show that after the collision P is moving with speed $\dfrac{u(3-e)}{4}$.

    b Show that the loss of kinetic energy due to the collision is $\dfrac{3mu^2(1-e^2)}{8}$.

    c Find in terms of m, u and e the impulse exerted on Q by P in the collision.

10 Two spheres of mass $70\,g$ and $100\,g$ are moving along a straight line towards each other with velocities $4\,m\,s^{-1}$ and $8\,m\,s^{-1}$ respectively. Their coefficient of restitution is $\frac{5}{12}$. Find their velocities after impact and the amount of kinetic energy lost in the collision.

11 A mass of $2\,kg$ moving at $35\,m\,s^{-1}$ catches up and collides with a mass of $10\,kg$ moving in the same direction at $20\,m\,s^{-1}$. Five seconds after the impact the $10\,kg$ mass encounters a fixed barrier which reduces it to rest. Assuming the coefficient of restitution between the masses is $\frac{3}{5}$, find the further time that will elapse before the $2\,kg$ mass strikes the $10\,kg$ mass again. You may assume that the masses are moving on a smooth surface and have constant velocity between collisions.

12 Three balls A, B and C of masses $4m$, $3m$ and $3m$, respectively and of equal radius lie at rest on a smooth horizontal table with their centres in a straight line. Their coefficient of restitution is $\frac{3}{4}$. Show that if A is projected towards B with speed V there are three impacts and the final velocities are $\frac{5}{32}V$, $\frac{1}{4}V$ and $\frac{7}{8}V$ respectively.

13 A bullet of mass $60\,g$ is fired horizontally at a fixed vertical metal barrier. The bullet hits the barrier when it is travelling at $600\,m\,s^{-1}$ and then rebounds.

    a Find the kinetic energy lost at the impact if $e = 0.4$.

    b Give one possible form of energy into which the lost kinetic energy has been transformed.

14 A particle A of mass $4m$ moving with speed u on a horizontal plane strikes directly a particle B of mass $3m$ which is at rest on the plane. The coefficient of restitution between A and B is e.

    a Find, in terms of e and u, the speeds of A and B immediately after the collision.

    b Given that the magnitude of the impulse exerted by A on B is $2mu$ show that $e = \frac{1}{6}$.

15 A ball of mass m moving with speed kV on a smooth table catches up and collides with another ball of mass λm moving with speed V travelling in the same direction on the table. The impact reduces the first ball to rest. Show that the coefficient of restitution is $\dfrac{\lambda + k}{\lambda(k-1)}$ and that λ must be greater than $\dfrac{k}{k-2}$ and k must be greater than 2.

16 A ball is dropped from zero velocity and after falling for $1\,s$ under gravity meets another equal ball which is moving upwards at $7\,m\,s^{-1}$.

    a Taking the value of g as $9.8\,m\,s^{-2}$, calculate the velocity of each ball immediately after the impact, given that the coefficient of restitution is $\frac{1}{4}$.

    b Find the percentage loss in kinetic energy due to the impact, giving your answer to 2 significant figures.

17 A particle falls from a height 8 m onto a fixed horizontal plane. The coefficient of restitution between the particle and the plane is $\frac{1}{4}$.

a Find the height to which the particle rises after impact.

b Find the time the particle takes from leaving the plane after impact to reach the plane again.

c What is the velocity of the particle after the second rebound?

You may leave your answers in terms of g.

18 A particle falls from a height h onto a fixed horizontal plane. If e is the coefficient of restitution between the particle and the plane, show that the total time taken before the particle finishes bouncing is $\frac{1+e}{1-e} \times \sqrt{\frac{2h}{g}}$.

19 A sphere P of mass m lies on a smooth table between a sphere Q of mass $8m$ and a fixed vertical plane. Sphere P is projected towards sphere Q. The coefficient of restitution between the two spheres is $\frac{7}{8}$. Given that sphere P is reduced to rest by the second impact with sphere Q, find the coefficient of restitution between sphere P and the fixed vertical plane.

20 A gun of mass M kg is free to move horizontally. The gun fires a shell of mass m kg in a horizontal direction. The energy released by the explosion, which occurred in firing the shell, is E J. Find the velocity of the shell in terms of m, M and E if all of this energy is given to the shell and the gun.

Summary of key points

1 The **impulse–momentum principle** states that the impulse of a force is equal to the change in momentum produced:

impulse = force × time

$\mathbf{I} = m\mathbf{v} - m\mathbf{u}$

where m is the mass of the body, $\mathbf{u}$ the initial velocity and $\mathbf{v}$ the final velocity.

2 The **principle of conservation of linear momentum** states that the total momentum before impact equals the total momentum after impact:

momentum = mass × velocity

$m_1\mathbf{u}_1 + m_2\mathbf{u}_2 = m_1\mathbf{v}_1 + m_2\mathbf{v}_2$

where a body of mass m_1 moving with velocity $\mathbf{u}_1$ collides with a body of mass m_2 moving with a velocity of $\mathbf{u}_2$. $\mathbf{v}_1$ and $\mathbf{v}_2$ are the velocities of the bodies after the collision.

3 **Newton's Law of Restitution** (also called Newton's experimental law) states that:

$$\frac{\text{speed of separation of particles}}{\text{speed of approach of particles}} = e$$

The constant e is the coefficient of restitution between the particles and $0 \leqslant e \leqslant 1$.

4 For the direct collision of a particle with a smooth plane Newton's Law of Restitution can be written as:

$$e = \frac{\text{speed of rebound}}{\text{speed of approach}}$$

$$e = \frac{v}{u}$$

You can also use

$v = eu$ or $u = \dfrac{v}{e}$

5 The loss of kinetic energy due to impact is:

$$\left(\tfrac{1}{2}m_1u_1{}^2 + \tfrac{1}{2}m_2u_2{}^2\right) - \left(\tfrac{1}{2}m_1v_1{}^2 + \tfrac{1}{2}m_2v_2{}^2\right)$$

Some kinetic energy is converted into heat or sound energy at impact, unless the impact is perfectly elastic.

After completing this chapter you should be able to

1 find the sum of the moments of the forces acting on a rigid body

2 solve problems about the equilibrium of a rigid body.

Statics of rigid bodies

5

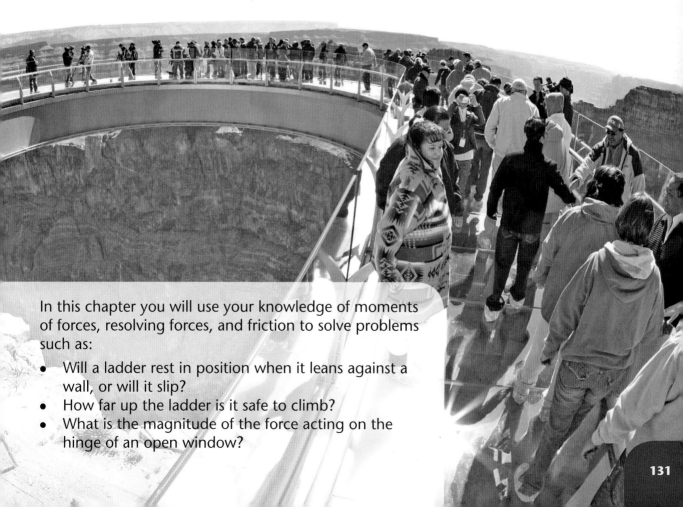

The Skywalk viewing platform hangs over the Grand Canyon (it is a *cantilevered* structure). The engineers would have needed to use moments to calculate the forces acting on the platform.

In this chapter you will use your knowledge of moments of forces, resolving forces, and friction to solve problems such as:

• Will a ladder rest in position when it leans against a wall, or will it slip?

• How far up the ladder is it safe to climb?

• What is the magnitude of the force acting on the hinge of an open window?

5.1 You can calculate the moment of a force acting on a body.

The moment of a force measures the turning effect of the force on the body on which it is acting.

■ **The moment of a force *F* about a point *P* is the product of the magnitude of the force and the perpendicular distance of the line of action of the force from the point *P*.**

The moment of the force is measured in newton-metres (Nm). When describing the turning effect of the force you need to consider its magnitude and the sense of the rotation (clockwise or anticlockwise).

Example 1

The diagram shows the forces acting on a lamina.
Find the sum of the moments of these forces about *P*.

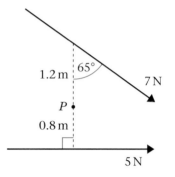

Method 1:

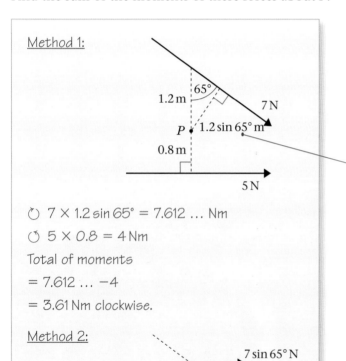

Because the line of action of the 7 N force is not perpendicular to the distance you have been given, you need to start by finding the perpendicular distance.

↻ $7 \times 1.2 \sin 65° = 7.612 \ldots$ Nm

↺ $5 \times 0.8 = 4$ Nm

Total of moments

$= 7.612 \ldots -4$

$= 3.61$ Nm clockwise.

Method 2:

Rather than find the perpendicular distance, you might find it easier to resolve the 7 N force into two components parallel and perpendicular to the given distance.

↻ $7 \sin 65° \times 1.2 = 7.612 \ldots$ Nm

↺ $5 \times 0.8 = 4$ Nm

Total of moments

$= 3.61$ Nm clockwise.

You get the same answer with both methods.

Exercise 5A

Find the sum of the moments about *P* of the forces shown in the following questions.

1
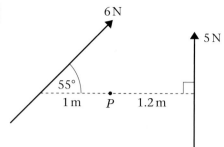

2
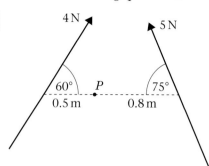

3
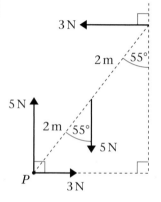

4
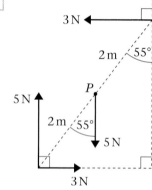

5
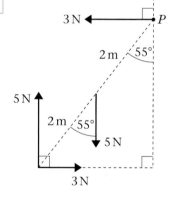

6
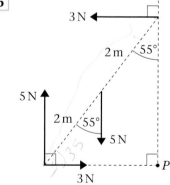

7
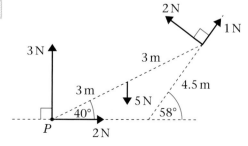

8
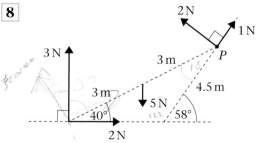

9

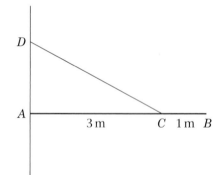

10

5.2 You can solve problems about rigid bodies resting in equilibrium.

■ If a body is resting in equilibrium then

● there is zero resultant force in any direction, i.e. the sum of the components of all the forces in any direction is zero,

● the sum of the moments about any point is zero.
NB The point does not need to be on the body.

Example 2

A uniform rod AB, of mass 6 kg and length 4 m, is smoothly hinged at A. A light inextensible string is attached to the rod at point C, where $AC = 3$ m, and to the point D, vertically above A. If the string keeps the rod in equilibrium in a horizontal position and the angle between the string and the rod is 40°, calculate

a the tension in the string,

b the magnitude and direction of the reaction at the hinge.

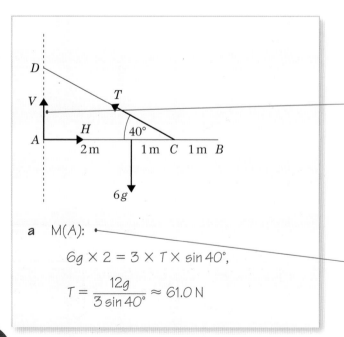

Start with a clear diagram showing all the forces acting.

It is usually simplest to start with the reaction at the hinge expressed in terms of its horizontal and vertical components.

The rod is resting in equilibrium, so the sum of the moments about any point is zero.

At present we do not know anything about the reaction at A. Since the reaction at A will have no turning effect about A, consider the sum of the moments of all of the forces about A.

b Consider all of the forces acting on AB:

$$R(\uparrow) \quad V + T\sin 40° = 6g, \ V = 19.6\,\text{N}$$
$$R(\rightarrow) \quad H = T\cos 40° = 46.7\,\text{N}$$
$$\Rightarrow \quad \text{The reaction has magnitude}$$
$$\sqrt{19.6^2 + 46.7^2} = 50.6\,\text{N}$$
$$\text{and acts at } \tan^{-1}\frac{19.6}{46.7} = 22.8° \text{ to } AB.$$

> The sum of the components of the forces in any direction is zero because AB is resting in equilibrium.

> Using Pythagoras' Theorem and trigonometry.

Alternative method

$$M(D)$$
$$6g \times 2 = H \times 3\tan 40°$$
$$H = 46.7\,\text{N}$$

$$M(C)$$
$$6g \times 1 = V \times 3$$
$$V = 2g = 19.6\,\text{N}$$

> We can find H without having to find T first, by taking moments about a point through which T and V pass.

> Similarly we can find V without needing T.

> Note that in both cases we have taken moments about a point which is *not* on the rod.

Example 3

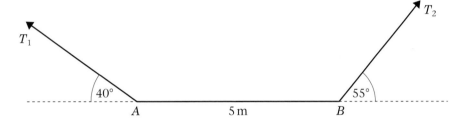

A non-uniform rod AB, of mass 3 kg and length 5 m, rests horizontally in equilibrium, supported by two strings attached at the ends of the rod. The strings make angles of 40° and 55° with the horizontal, as shown in the diagram.

a Find the magnitudes of the tensions in the two strings.

b Find the distance of the centre of mass of the rod from A.

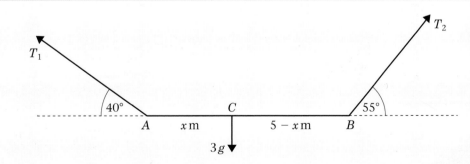

Suppose that the centre of mass of AB is at C, a distance x m from A.

a $R(\rightarrow) \quad T_1\cos 40° = T_2\cos 55°$

$\quad R(\uparrow) \quad 3g = T_1\sin 40° + T_2\sin 55°$

> Start with a clear diagram showing the things you know and the things you need to find.

$$\Rightarrow T_2 = \frac{T_1 \cos 40°}{\cos 55°}$$

Use the first equation to express T_2 in terms of T_1, then substitute this expression into the second equation to obtain an equation with just one unknown value.

$$3g = T_1 \sin 40° + \frac{T_1 \cos 40°}{\cos 55°} \sin 55°$$

$$3g \approx T_1 \times 1.7368$$

$$T_1 \approx 16.9\,\text{N},\ T_2 \approx 22.6\,\text{N}$$

b M(A):

$$x \times 3g = 5 \times T_2 \times \sin 55°$$

Use moments to obtain an equation in x.

$$x = \frac{5T_2 \sin 55°}{3g} \approx 3.1$$

Exercise 5B

Whenever a numerical value of g is required, take $g = 9.8\,\text{m s}^{-2}$.

1 Each of the following diagrams shows a uniform beam AB of length 4 m and mass 6 kg. The beam is freely hinged at A and resting horizontally in equilibrium. In each case find

i the magnitude of the force T,

ii the magnitude and direction of the reaction at A.

a

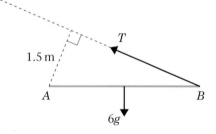

b

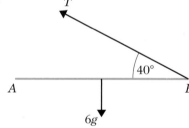

c

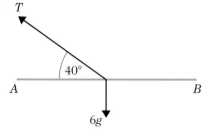

d

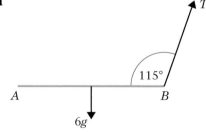

2 Each of the following diagrams shows a uniform rod XY of mass $4\,\text{kg}$ and length $5\,\text{m}$. The rod is freely hinged to a vertical wall at X. The rod rests in equilibrium at an angle to the horizontal. Find the magnitude of the force F in each case.

a

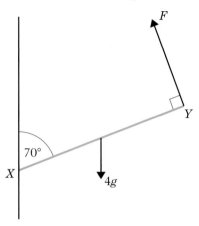

b

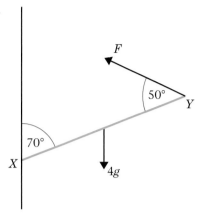

c

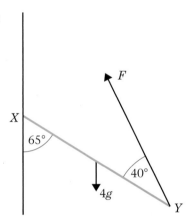

d

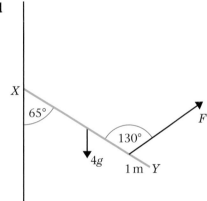

3 A uniform rod AB of length $2a\,\text{m}$ and mass $m\,\text{kg}$ is smoothly hinged at A. It is maintained in equilibrium by a horizontal force of magnitude P acting at B. The rod is inclined at $30°$ to the horizontal with B below A.

a Show that $P = \dfrac{\sqrt{3}}{2}mg$.

b Find the magnitude and direction of the reaction at the hinge.

4 A uniform beam AB of mass $10\,\text{kg}$ and length $3\,\text{m}$ is attached to a vertical wall by means of a smooth hinge at A. The beam is maintained in the horizontal position by means of a light inextensible string, one end of which is attached to the beam at B and the other end of which is attached to the wall at a point $2\,\text{m}$ vertically above A.

a Find the tension in the string.

A particle of mass $M\,\text{kg}$ is now attached to the beam at B.

b Given that the tension in the string is now double its original value, find the value of M.

5 A uniform horizontal beam AB of mass 5 kg is freely hinged to a vertical wall and is supported by a rod CD as shown in the diagram. Given that the tension in the rod is 80 N, $AC = 1$ m and the angle between the rod and the vertical is 45°, find the length of the beam.

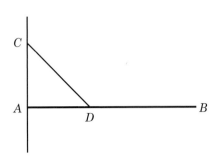

6 $ABCD$ is a uniform rectangular lamina with mass 5 kg, side $AB = 1$ m, and side $AD = 2$ m. It is hinged at A so that it is free to move in a vertical plane. It is maintained in equilibrium, with B vertically below A, by a horizontal force acting at C and a vertical force acting at D, each of magnitude F N. Find

a the value of F,

b the magnitude and direction of the force exerted by the hinge on the lamina.

7

A uniform rod AB of mass 3 kg and length 2 m rests horizontally in equilibrium supported by two strings attached at the ends of the rod. The strings make angles of 40° and θ with the horizontal, as shown in the diagram. Find the magnitudes of the tensions in the strings and the value of θ.

8 A non-uniform horizontal beam AB of mass 5 kg and length 3 m is freely hinged to a vertical wall and is supported by a rod CD as shown in the diagram. Given that the thrust in the rod is 35 N, $AC = 1$ m and the angle between the rod and the vertical is 45°, find the distance of the centre of mass of the beam from A.

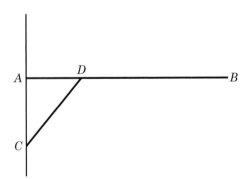

9 $PQRS$ is a uniform square lamina of side 3 m and mass 6 kg. It is freely hinged at P so that it is free to move in a vertical plane. It is maintained in equilibrium, with PR horizontal, and Q above S, by a force of magnitude F N acting along SR and a force of magnitude $2F$ N acting along RQ. Find

a the value of F,

b the magnitude and direction of the force exerted by the hinge on the lamina.

10

A non-uniform rod *AB* of mass 5 kg and length 4 m, with a particle of mass 2 kg attached at *B*, rests horizontally in equilibrium supported by two strings attached at the ends of the rod. The strings make angles of 30° and 45° with the horizontal, as shown in the diagram. Find

a the tensions in each of the strings,

b the position of the centre of mass of the rod.

> **5.3** **When a rigid body is resting in equilibrium under the action of three non-parallel forces you can solve problems using a triangle of forces.**

Example 4

A non-uniform rod *AB*, of mass 5 kg and length 4 m, rests horizontally in equilibrium supported by two strings attached at the ends of the rod. The strings make angles of 30° and 60° with the horizontal, as shown in the diagram. Find

a the position of the centre of mass of the rod,

b the tensions in the two strings.

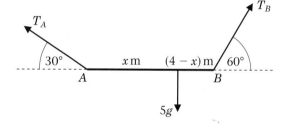

a

Suppose that the centre of mass is x m from *A*. Consider the moments of the forces about the point where the lines of action of T_A and T_B meet. The sum of the moments of T_A and T_B about this point is zero. Since the rod is resting in equilibrium, we know that the sum of the moments of all three forces about any point is zero. It follows that the moment of the weight of the rod about this point must be zero and therefore the line of action of this force must pass through the point.

The vertical distance from the rod to the point can be expressed in 2 ways: $x \tan 30°$ and $(4 - x) \tan 60°$.

Therefore $x \times \dfrac{1}{\sqrt{3}} = (4 - x) \times \sqrt{3}$

$x = (4 - x) \times 3$

$4x = 12, x = 3$ m

So the centre of mass lies 3 m from *A*.

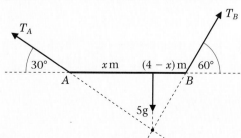

b

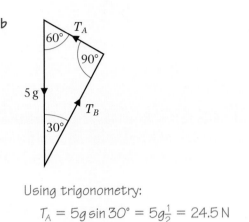

Using trigonometry:

$$T_A = 5g \sin 30° = 5g\frac{1}{2} = 24.5\,\text{N}$$

$$\text{and } T_B = 5g \sin 60° = 5g\frac{\sqrt{3}}{2} = 42.4\,\text{N}$$

Since the rod is resting in equilibrium, the vector sum of the three forces acting must be zero, so you can draw a triangle of forces.

■ **If a rigid body is resting in equilibrium under the action of just three non-parallel forces**
  ● **the lines of action of the three forces are concurrent (i.e. they all pass through a common point)**
  ● **you can draw a triangle of vectors to represent the three forces.**

Exercise 5C

Whenever a numerical value of g is required, take $g = 9.8\,\text{m s}^{-2}$. Use the method described in Example 4 to answer the questions in this exercise.

1 A non-uniform rod AB, of mass 4 kg and length 4 m, rests horizontally in equilibrium supported by two strings attached at the ends of the rod. The strings make angles of 30° and 50° with the horizontal, as shown in the diagram. Find

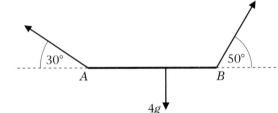

  a the position of the centre of mass of the rod,

  b the tensions in the two strings.

2 A uniform rod AB of mass 5 kg and length 3 m is freely hinged to a vertical wall at A. The rod is maintained in horizontal equilibrium by a force P N acting at B, as shown in the diagram. Find

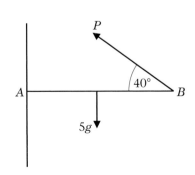

  a the magnitude of P,

  b the magnitude and direction of the reaction of the force exerted by the hinge on the rod.

3 A window of mass 15 kg and height 120 cm is hinged along its top edge. It is kept open by the thrust from a light strut of length 50 cm attached to the wall and perpendicular to the lower edge of the window. By modelling the window as a uniform lamina, calculate the thrust in the strut and the magnitude and direction of the force exerted on the window by the hinge.

4 A uniform rod AB of weight 20 N and length 3 m is freely hinged to a vertical wall at A. A force P is applied at B at right angles to the rod in order to keep the rod in equilibrium at an angle of 30° to the horizontal with B above A. Find

a the magnitude of P,

b the magnitude and direction of the reaction at the hinge.

5 AB is a loaded shelf freely hinged to the wall at A, and supported in a horizontal position by a light strut CD, which is attached to the shelf at D, 25 cm from A, and attached to the wall at C, 25 cm below A. The total weight of the shelf and its load is 80 N, and the centre of mass is 20 cm from A. Find the thrust in the strut and the magnitude and direction of the force exerted by the hinge on the shelf.

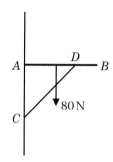

6 A uniform rod XY of mass 6 kg is suspended from the ceiling by light inextensible strings attached to its ends. The rod is resting in equilibrium at 10° to the horizontal with X below Y. The string attached to Y is at an angle of 35° to the vertical, and is fixed to the ceiling at a point above the rod. Find

a the angle that the other string makes with the vertical,

b the tensions in the two strings.

5.4 You can solve problems about rigid bodies resting in limiting equilibrium.

When a body is resting in equilibrium and one of the forces acting on the body is a frictional force, you need to consider whether the body is on the point of moving.

■ **If a body is on the point of moving then it is said to be in limiting equilibrium. In this case the frictional force takes its maximum value, μR, where μ is the coefficient of friction and R is the normal reaction.**

Example **5**

A uniform rod AB of mass 40 kg and length 10 m rests with the end A on rough horizontal ground. The rod rests against a smooth peg C where $AC = 8$ m. The rod is in limiting equilibrium at an angle of 15° to the horizontal. Find

a the magnitude of the reaction at C,

b the coefficient of friction between the rod and the ground.

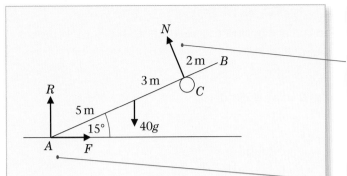

Start with a diagram showing all the forces.

N, the reaction at C, is perpendicular to the rod. The peg is smooth, so there is no friction here.

At A there is a normal reaction and a frictional force.

a $40g \cos 15° \times 5 = N \times 8$

$N = \dfrac{40g \cos 15° \times 5}{8} = 25g \cos 15°$

$= 236.65 \dots N$

Taking moments about A.

The reaction at C has magnitude $237 N$ (3 s.f.).

b $R(\rightarrow)$ $F = N \cos 75° = 61.25 \dots N$

$R(\uparrow)$ $R + N \cos 15° = 40g$

$R = 40g - N \cos 15° = 163.4 N$

The rod is in limiting equilibrium, so

$F = \mu R, \ \mu = \dfrac{F}{R} = 0.37$

Resolving horizontally and vertically.

Example 6

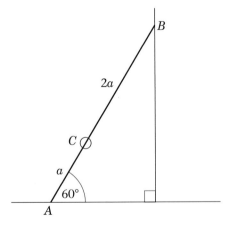

A ladder AB, of mass m and length $3a$, has one end A resting on rough horizontal ground. The other end B rests against a smooth vertical wall. A load of mass $2m$ is fixed on the ladder at the point C, where $AC = a$. The ladder is modelled as a uniform rod in a vertical plane perpendicular to the wall and the load is modelled as a particle. The ladder rests in limiting equilibrium at an angle of $60°$ with the ground.

Find the coefficient of friction between the ladder and the ground.

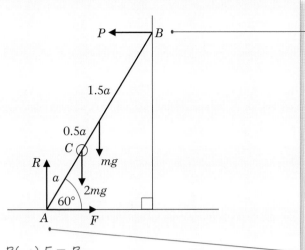

The reaction at B is perpendicular to the wall. The wall is smooth, so there is no friction at B.

$R(\rightarrow) \quad F = P$

$R(\uparrow) \quad R = 2mg + mg = 3mg$

Consider the point where the lines of action of R and P meet. Take moments about this point:

$$\frac{7mga}{2} \times \frac{1}{2} = F \times 3a\frac{\sqrt{3}}{2}$$

$$F = \frac{7mga \times 2}{4 \times 3a\sqrt{3}} = \frac{7mg}{6\sqrt{3}}$$

The ladder is in limiting equilibrium, so

$$F = \mu R$$

$$\mu = \frac{F}{R} = \frac{\left(\dfrac{7mg}{6\sqrt{3}}\right)}{3mg} = \frac{7}{18\sqrt{3}} = 0.22$$

At A there is a normal reaction and friction.

Resolving horizontally and vertically.

There are several points that you could choose to take moments about, but by choosing the point where the lines of action of R and P meet we obtain an equation in F only. This makes the working easier.

Exercise 5D

Whenever a numerical value of g is required, take $g = 9.8 \text{ m s}^{-2}$.

1. A uniform rod AB of weight 80 N rests with its lower end A on a rough horizontal floor. A string attached to end B keeps the rod in equilibrium. The string is held at 90° to the rod. The tension in the string is T. The coefficient of friction between the rod and the ground is μ. R is the normal reaction at A and F is the frictional force at A. Find the magnitudes of T, R and F, and the least possible value of μ.

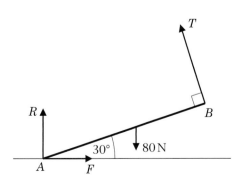

2 A uniform ladder of mass 10 kg and length 5 m rests against a smooth vertical wall with its lower end on rough horizontal ground. The ladder rests in equilibrium at an angle of 65° to the horizontal. Find

  a the magnitude of the normal contact force S at the wall,

  b the magnitude of the normal contact force R at the ground and the frictional force at the ground,

  c the least possible value of the coefficient of friction between the ladder and the ground.

3 A uniform ladder AB of mass 20 kg rests with its top A against a smooth vertical wall and its base B on rough horizontal ground. The coefficient of friction between the ladder and the ground is $\frac{3}{4}$. A mass of 10 kg is attached to the ladder. Given that the ladder is about to slip, find the inclination of the ladder to the horizontal

  a if the 10 kg mass is attached at A,

  b if the 10 kg mass is attached at B.

4 A uniform ladder of mass 20 kg and length 8 m rests against a smooth vertical wall with its lower end on rough horizontal ground. The coefficient of friction between the ground and the ladder is 0.3. The ladder is inclined at an angle θ to the horizontal, where $\tan \theta = 2$. A boy of mass 30 kg climbs up the ladder. Find how far up the ladder he can climb without it slipping.

5 A smooth horizontal rail is fixed at a height of 3 m above a rough horizontal surface. A uniform pole AB of weight 4 N and length 6 m is resting with end A on the rough ground and touching the rail at point C. The vertical plane containing the pole is perpendicular to the rail. The distance AC is 4.5 m and the pole is in limiting equilibrium. Calculate

  a the magnitude of the force exerted by the rail on the pole,

  b the coefficient of friction between the pole and the ground.

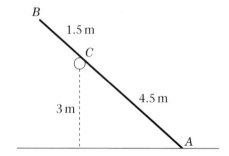

6 A uniform ladder rests in limiting equilibrium with its top against a smooth vertical wall and its base on a rough horizontal floor. The coefficient of friction between the ladder and the floor is μ. Given that the ladder makes an angle θ with the floor, show that $2\mu \tan \theta = 1$.

7

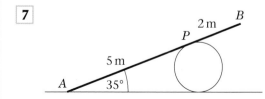

A uniform ladder AB has length 7 m and mass 20 kg. The ladder is resting against a smooth cylindrical drum at P, where AP is 5 m, with end A in contact with rough horizontal ground. The ladder is inclined at 35° to the horizontal.

Find the normal and frictional components of the contact force at A, and hence find the least possible value of the coefficient of friction between the ladder and the ground.

8 A uniform ladder rests in limiting equilibrium with one end on rough horizontal ground and the other end against a rough vertical wall. The coefficient of friction between the ladder and the ground is μ_1 and the coefficient of friction between the ladder and the wall is μ_2. Given that the ladder makes an angle θ with the horizontal, show that $\tan \theta = \dfrac{1 - \mu_1\mu_2}{2\mu_1}$.

9 A uniform ladder of weight W rests in equilibrium with one end on rough horizontal ground and the other resting against a smooth vertical wall. The vertical plane containing the ladder is at right angles to the wall and the ladder is inclined at $60°$ to the horizontal. The coefficient of friction between the ladder and the ground is μ.

a Find, in terms of W, the magnitude of the force exerted by the wall on the ladder.

b Show that $\mu \geqslant \frac{1}{6}\sqrt{3}$.

A load of weight w is attached to the ladder at its upper end (resting against the wall).

c Given that $\mu = \frac{1}{5}\sqrt{3}$ and that the equilibrium is limiting, find w in terms of W.

10

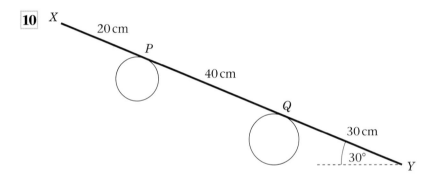

A uniform rod XY has weight $20\,\text{N}$ and length $90\,\text{cm}$. The rod rests on two parallel pegs, with X above Y, in a vertical plane which is perpendicular to the axes of the pegs, as shown in the diagram. The rod makes an angle of $30°$ to the horizontal and touches the two pegs at P and Q, where $XP = 20\,\text{cm}$ and $XQ = 60\,\text{cm}$.

a Calculate the normal components of the forces on the rod at P and at Q.

The coefficient of friction between the rod and each peg is μ.

b Given that the rod is about to slip, find μ.

11 The diagram shows the vertical cross section $ABCD$ through the centre of mass of a uniform rectangular box. The box is resting on a rough horizontal floor and leaning against a smooth vertical wall. The box has mass $25\,\text{kg}$. $AB = 0.5\,\text{m}$, $BC = 1.5\,\text{m}$ and AD is at an angle of θ to the horizontal. The coefficient of friction between the box and the ground is $\frac{1}{4}$. Given that the box is about to slip, find the value of θ.

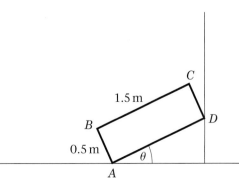

12 A uniform rod of length *l* has a ring at one end which can slide along a rough horizontal pole. The coefficient of friction between the ring and the pole is 0.2. The other end of the rod is attached to the end of the pole by a light inextensible cord of length *l*. The rod rests in equilibrium at an angle of θ to the horizontal. Using a geometrical method, or otherwise, find the smallest possible value of θ.

Mixed exercise 5E

Whenever a numerical value of *g* is required, take $g = 9.8 \text{ m s}^{-2}$.

1 A uniform ladder of mass 20 kg and length 6 m rests with one end on a smooth horizontal floor and the other end against a smooth vertical wall. The ladder is held in this position by a light inextensible rope of length 5 m which has one end attached to the bottom of the ladder and the other end fastened to a point at the base of the wall, vertically below the top of the ladder. Find the tension in the rope.

2 A uniform ladder *AB* of mass *M* kg and length 5 m rests with end *A* on a smooth horizontal floor and end *B* against a smooth vertical wall. The ladder is held in equilibrium at an angle θ to the floor by a light horizontal string attached to the wall and to a point *C* on the ladder. If $\tan \theta = 2$, find the tension in the string when the length *AC* is 2 m.

3 A uniform pole *AB* of weight 12 N has its lower end *A* on rough horizontal ground. The pole is being raised into a vertical position by a rope attached to *B*. The rope and the pole lie in the same vertical plane and *A* does not slip across the ground. Find the horizontal and vertical components of the reaction at the ground when the rope is perpendicular to the pole and the pole is at 15° to the horizontal.

4 *AB* is a light rod of length 5*a* rigidly joined to a light rod *BC* of length 2*a* so that the rods are perpendicular to each other and in the same vertical plane, as shown in the diagram. The centre, *O*, of *AB* is fixed and the rods can rotate freely about *O* in a vertical plane. A particle of mass 3*m* is attached at *A* and a particle of mass *m* is attached at *C*. The system rests in equilibrium with *AB* inclined at an acute angle θ to the vertical. Find the value of θ.

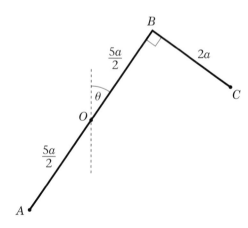

5 A uniform ladder *AB* has one end *A* on smooth horizontal ground. The other end *B* rests against a smooth vertical wall. The ladder is modelled as a uniform rod of mass *m* and length 5*a*. The ladder is kept in equilibrium by a horizontal force *F* acting at a point *C* of the ladder where *AC = a*. The force *F* and the ladder lie in a vertical plane perpendicular to the wall. The ladder is inclined to the horizontal at an angle θ, where $\tan \theta = 1.8$, as shown in the diagram.

Show that $F = \dfrac{25mg}{72}$.

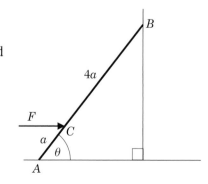

6 A uniform rod AB, of length $12a$ and weight W, is free to rotate in a vertical plane about a smooth pivot at A. One end of a light inextensible string is attached to B. The other end is attached to point C which is vertically above A, with $AC = 5a$. The rod is in equilibrium with AB horizontal, as shown in the diagram. Find, in terms of W,

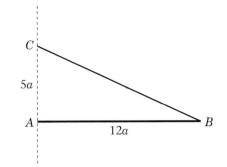

a the tension in the string,

b the magnitude of the horizontal component of the force exerted by the pivot on the rod.

7 A uniform ladder AB, of mass m and length $2a$, has one end A on rough horizontal ground. The other end B rests against a smooth vertical wall. The ladder is in a vertical plane perpendicular to the wall. The ladder makes an angle α with the vertical, where $\tan \alpha = \frac{3}{4}$. A child of mass $2m$ stands on the ladder at C where $AC = \frac{2}{3}a$, as shown in the diagram. The ladder and the child are in equilibrium.

By modelling the ladder as a rod and the child as a particle, calculate the least possible value of the coefficient of friction between the ladder and the ground.

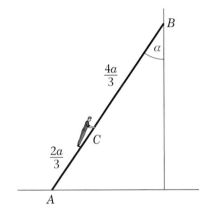

8 A uniform steel girder AB of weight $400\,\text{N}$ and length $4\,\text{m}$, is freely hinged at A to a vertical wall. The girder is supported in a horizontal position by a steel cable attached to the girder at B. The other end of the cable is attached to the point C vertically above A on the wall, with $\angle ABC = \alpha$ where $\tan \alpha = \frac{1}{2}$. A load of weight $200\,\text{N}$ is suspended by another cable from the girder at the point D, where $AD = 3\,\text{m}$, as shown in the diagram. The girder remains horizontal and in equilibrium. The girder is modelled as a rod, and the cables as light inextensible strings.

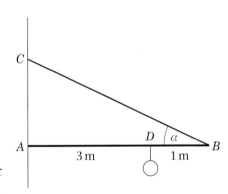

a Show that the tension in the cable BC is $350\sqrt{5}\,\text{N}$.

b Find the magnitude of the reaction on the girder at A.

9 A non-uniform rod AB of length l rests horizontally with its ends resting on two smooth surfaces inclined at $20°$ and $30°$ to the horizontal, as shown in the diagram. Use a geometrical method to find the distance of the centre of mass from A.

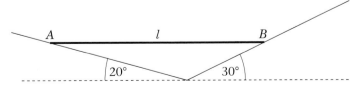

10 A uniform ladder, of weight W and length $2a$, rests in equilibrium with one end A on a smooth horizontal floor and the other end B against a rough vertical wall. The ladder is in a vertical plane perpendicular to the wall. The coefficient of friction between the wall and the ladder is μ. The ladder makes an angle θ with the floor, where $\tan \theta = \frac{4}{3}$. A horizontal light inextensible string CD is attached to the ladder at the point C, where $AC = \frac{1}{4}a$. The string is attached to the wall at the point D, with BD vertical, as shown in the diagram. The tension in the string is $\frac{1}{3}W$. By modelling the ladder as a rod,

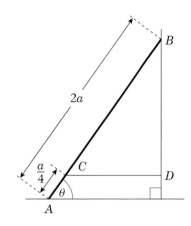

a find the magnitude of the force of the floor on the ladder,

b show that $\mu \geqslant \frac{1}{3}$.

c State how you have used the modelling assumption that the ladder is a rod.

11 A uniform pole AB of mass $40\,\text{kg}$ and length $3\,\text{m}$, is smoothly hinged to a vertical wall at one end A. The pole is held in equilibrium in a horizontal position by a light rod CD. One end C of the rod is fixed to the wall vertically below A. The other end D is freely jointed to the pole so that $\angle ACD = 45°$ and $AC = 1\,\text{m}$, as shown in the diagram. Find

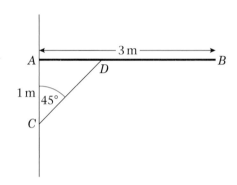

a the thrust in the rod CD,

b the magnitude of the force exerted by the wall on the pole at A.

The rod CD is removed and replaced by a longer light rod CM, where M is the mid-point of AB. The rod is freely jointed to the pole at M. The pole AB remains in equilibrium in a horizontal position.

c Show that the force exerted by the wall on the pole at A now acts horizontally.

12 A uniform rod AB, of weight W and length $2a$, is used to display a light banner. The rod is freely hinged to a vertical wall at point B. It is held in a horizontal position by a light wire attached to A and to a point C vertically above B on the wall. The angle CAB is θ, where $\tan \theta = \frac{1}{3}$.

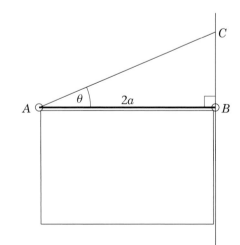

a Show that the tension in the wire is

$$\frac{W}{2 \sin \theta}.$$

b Find, in terms of W, the magnitude of the force exerted by the wall on the rod at B.

13 A uniform ladder, of weight W and length 5 m, has one end on rough horizontal ground and the other touching a smooth vertical wall. The coefficient of friction between the ladder and the ground is 0.3.

Given that the top of the ladder touches the wall at a point 4 m vertically above the level of the ground,

a show that the ladder can not rest in equilibrium in this position.

In order to enable the ladder to rest in equilibrium in the position described above, a brick is attached to the bottom of the ladder.

Assuming that this brick is at the lowest point of the ladder, but not touching the ground,

b show that the horizontal frictional force exerted by the ladder on the ground is independent of the mass of the brick,

c find, in terms of W and g, the smallest mass of the brick for which the ladder will rest in equilibrium.

14

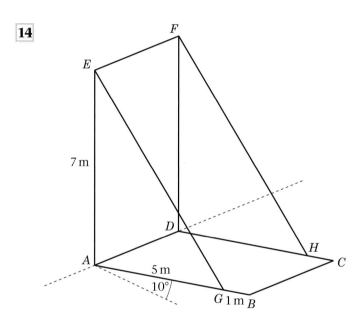

A uniform drawbridge $ABCD$ is 6 m long and has mass 1200 kg. The bridge is held at 10° to the horizontal by two chains attached to points G and H on the bridge 5 m from the hinge and to fixed points E and F at a height of 7 m vertically above A and D. Find the force from the hinge.

Summary of key points

1 The moment of a force F about a point P is the product of the magnitude of the force and the perpendicular distance of the line of action of the force from the point P.
The moment of the force is measured in newton-metres (Nm).

2 When describing the turning effect of the force you need to consider its magnitude and the sense of the rotation (clockwise or anticlockwise).

3 Taking moments about a point through which an unknown or unwanted force passes eliminates that force, and can simplify a problem.

4 When a rigid body is in equilibrium then
- the sum of the components of all the forces acting on the body in any direction is zero,
- the sum of the moments of all the forces acting on the body about any point is zero. NB The point does not have to be a point on the body.

5 If a rigid body is resting in equilibrium under the action of just three non-parallel forces
- the lines of action of the three forces are concurrent (i.e. they all pass through a common point)
- you can draw a triangle of vectors to represent the three forces

6 If a body is on the point of moving then it is said to be in limiting equilibrium. The frictional force takes its maximum value, μR, where μ is the coefficient of friction, and R is the normal reaction.

7 The normal reaction is perpendicular to the surface. In the case of a rod resting on a sphere, cylinder, bar, or peg, this means that the reaction is perpendicular to the rod (which lies on a tangent to the curved surface).

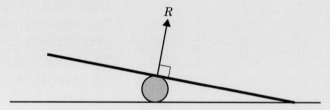

Review Exercise

Whenever a numerical value of g is required, take $g = 9.8\,\text{m s}^{-2}$.

1 [*In this question* **i** *and* **j** *are perpendicular unit vectors in a horizontal plane.*]

A ball has a mass 0.2 kg. It is moving with velocity $(30\mathbf{i})\,\text{m s}^{-1}$ when it is struck by a bat. The bat exerts an impulse of $(-4\mathbf{i} + 4\mathbf{j})\,\text{Ns}$ on the ball. Find

a the velocity of the ball immediately after the impact,

b the angle through which the ball is deflected as a result of the impact,

c the kinetic energy lost by the ball in the impact. **E**

2 A particle P of mass 0.75 kg is moving under the action of a single force **F** newtons. At time t seconds, the velocity $\mathbf{v}\,\text{m s}^{-1}$ of P is given by

$$\mathbf{v} = (t^2 + 2)\mathbf{i} - 6t\mathbf{j}.$$

a Find the magnitude of **F** when $t = 4$.

When $t = 5$, the particle P receives an impulse of magnitude $9\sqrt{2}\,\text{Ns}$ in the direction of the vector $\mathbf{i} - \mathbf{j}$.

b Find the velocity of P immediately after the impulse. **E**

3 A tennis ball of mass 0.2 kg is moving with velocity $(-10\mathbf{i})\,\text{m s}^{-1}$ when it is struck by a tennis racket. Immediately after being struck, the ball has velocity $(15\mathbf{i} + 15\mathbf{j})\,\text{m s}^{-1}$. Find

a the magnitude of the impulse exerted by the racket on the ball,

b the angle, to the nearest degree, between the vector **i** and the impulse exerted by the racket,

c the kinetic energy gained by the ball as a result of being struck. **E**

4 At time t seconds the acceleration, $\mathbf{a}\,\text{m s}^{-2}$, of a particle P relative to a fixed origin O, is given by $\mathbf{a} = 2\mathbf{i} + 6t\mathbf{j}$. Initially the velocity of P is $(2\mathbf{i} - 4\mathbf{j})\,\text{m s}^{-1}$.

a Find the velocity of P at time t seconds.

At time $t = 2$ seconds the particle P is given an impulse $(3\mathbf{i} - 1.5\mathbf{j})\,\text{Ns}$.

Given that the particle P has mass 0.5 kg,

b find the speed of P immediately after the impulse has been applied. **E**

5 The unit vectors **i** and **j** lie in a vertical plane, **i** being horizontal and **j** vertical. A ball of mass 0.1 kg is hit by a bat which gives it an impulse of $(3.5\mathbf{i} + 3\mathbf{j})$ Ns. The velocity of the ball immediately after being hit is $(10\mathbf{i} + 25\mathbf{j})$ m s^{-1}.

  a Find the velocity of the ball immediately before it is hit.

In the subsequent motion the ball is modelled as a particle moving freely under gravity. When it is hit the ball is 1 m above horizontal ground.

  b Find the greatest height of the ball above the ground in the subsequent motion.

The ball is caught when it is again 1 m above the ground.

  c Find the distance from the point where the ball is hit to the point where it is caught. **E**

6 Two particles, A and B, of mass m and $3m$ respectively, lie at rest on a smooth horizontal table. The coefficient of restitution between the particles is 0.25. The particles A and B are given speeds of $7u$ and u respectively towards each other so that they collide directly. Find

  a the speeds of A and B after the collision,

  b the loss in kinetic energy due to the collision. **E**

7 Two uniform smooth spheres A and B are of equal size and have masses $3m$ and $2m$ respectively. They are both moving in the same straight line with speed u, but in opposite directions, when they are in direct collision with each other. Given that A is brought to rest by the collision, find

  a the coefficient of restitution between the spheres,

  b the kinetic energy lost in the impact. **E**

8 A smooth sphere S of mass m is moving on a smooth horizontal plane with speed u. It collides directly with another smooth sphere T, of mass $3m$, whose radius is the same as S. The sphere T is moving in the same direction as S with speed $\frac{1}{6}u$. The sphere S is brought to rest by the impact. Find the coefficient of restitution between the spheres. **E**

9 A smooth sphere S of mass m is moving with speed u on a smooth horizontal plane. The sphere S collides with another smooth sphere T, of equal radius to S but of mass km, moving in the same straight line and in the same direction with speed λu, $0 < \lambda < \frac{1}{2}$. The coefficient of restitution between S and T is e.

Given that S is brought to rest by the impact,

  a show that $e = \dfrac{1 + k\lambda}{k(1 - \lambda)}$.

  b Deduce that $k > 1$. **E**

10 A particle A, of mass $2m$, is moving with speed u on a horizontal table when it collides directly with a particle B, of mass $3m$, which is at rest. The coefficient of restitution between the particles is e.

  a Find, in terms of e and u, the velocities of A and B immediately after the collision.

  b Show that, for all possible values of e, the speed of A immediately after the collision is not greater than $\frac{2}{5}u$.

Given that the magnitude of the impulse exerted by B on A is $\frac{11}{5}mu$,

  c find the value of e. **E**

11 A sphere P, of mass m, is moving in a straight line with speed u on the surface of a smooth horizontal table. Another sphere Q, of mass $5m$ and having the same radius as P, is initially at rest on the table. The sphere P strikes the sphere Q directly, and the direction of motion of P is reversed by the impact. The coefficient of restitution between P and Q is e.

a Find an expression, in terms of u and e, for the speed of P after the impact.

b Find the set of possible values of e. **E**

12 A smooth sphere A of mass m is moving with speed u on a smooth horizontal table when it collides directly with another smooth sphere B of mass $3m$, which is at rest on the table. The coefficient of restitution between A and B is e. The spheres have the same radius and are modelled as particles.

a Show that the speed of B immediately after the collision is $\frac{1}{4}(1 + e)u$.

b Find the speed of A immediately after the collision.

Immediately after the collision the total kinetic energy of the spheres is $\frac{1}{6}mu^2$.

c Find the value of e.

d Hence show that A is at rest after the collision.

13 A particle P of mass m is moving with speed $3u$ in a straight line on a smooth horizontal plane. It collides with another particle Q of mass $2m$ which is moving with speed $2u$ along the same straight line but in the opposite direction. The coefficient of restitution between P and Q is e. The magnitude of the impulse given to each particle during the collision is $5mu$, and both P and Q have their directions of motion reversed by the collision.

a Show that $e = \frac{1}{2}$.

b Calculate the loss of kinetic energy due to the collision. **E**

14 A smooth uniform sphere S of mass m is moving on a smooth horizontal plane with speed u. The sphere collides directly with another smooth uniform sphere T, of the same radius as S and a mass $2m$, which is at rest on the plane. The coefficient of restitution between the spheres is e.

a Show that the speed of T after the collision is $\frac{1}{3}u(1 + e)$.

Given that $e > \frac{1}{2}$,

b **i** find the speed of S after the collision,

ii determine whether the direction of motion of S is reversed by the collision. **E**

15 A particle P of mass $3m$ is moving with speed $2u$ in a straight line on a smooth horizontal table. The particle P collides with a particle Q of mass $2m$ moving with speed u in the opposite direction to P. The coefficient of restitution between P and Q is e.

a Show that the speed of Q after the collision is $\frac{1}{5}u(9e + 4)$.

As a result of the collision, the direction of motion of P is reversed.

b Find the range of possible values of e.

Given that the magnitude of the impulse of P on Q is $\frac{32}{5}mu$,

c find the value of e. **E**

16 A small smooth ball A of mass m is moving on a horizontal table with speed u when it collides directly with another small smooth ball B of mass $3m$ which is at rest on the table. The balls have the same radius and the coefficient of restitution between the balls is e. The direction of motion of A is reversed as a result of the collision.

a Find, in terms of e and u, the speeds of A and B immediately after the collision.

In the subsequent motion B strikes a vertical wall, which is perpendicular to the direction of motion of B, and rebounds. The coefficient of restitution between B and the wall is $\frac{3}{4}$.

Given that there is a second collision between A and B,

b find the range of values of *e* for which the motion described is possible. **(E)**

17 Two small spheres A and B have mass 3*m* and 2*m* respectively. They are moving towards each other in opposite directions on a smooth horizontal plane, both with speed 2*u*, when they collide directly. As a result of the collision, the direction of motion of B is reversed and its speed is unchanged.

a Find the coefficient of restitution between the spheres.

Subsequently, B collides directly with another small sphere C of mass 5*m* which is at rest. The coefficient of restitution between B and C is $\frac{3}{5}$.

b Show that, after B collides with C, there will be no further collisions between the spheres. **(E)**

18 A smooth sphere P of mass 2*m* is moving in a straight line with speed *u* on a smooth horizontal table. Another smooth sphere Q of mass *m* is at rest on the table. The sphere P collides directly with Q. The coefficient of restitution between P and Q is $\frac{1}{3}$. The spheres are modelled as particles.

a Show that, immediately after the collision, the speeds of P and Q are $\frac{5}{9}u$ and $\frac{8}{9}u$ respectively.

After the collision, Q strikes a fixed vertical wall which is perpendicular to the direction of motion of P and Q. The coefficient of restitution between Q and the wall is *e*. When P and Q collide again, P is brought to rest.

b Find the value of *e*.

c Explain why there must be a third collision between P and Q. **(E)**

19 Two small smooth spheres, P and Q, of equal radius, have masses 2*m* and 3*m* respectively. The sphere P is moving with speed 5*u* on a smooth horizontal table when it collides directly with Q, which is at rest on the table. The coefficient of restitution between P and Q is *e*.

a Show that the speed of Q immediately after the collision is 2(1 + *e*)*u*.

After the collision, Q hits a smooth vertical wall which is at the edge of the table and perpendicular to the direction of motion of Q. The coefficient of restitution between Q and the wall is *f*, 0 < *f* ≤ 1.

b Show that, when *e* = 0.4, there is a second collision between P and Q.

Given that *e* = 0.8 and there is a second collision between P and Q,

c find the range of possible values of *f*. **(E)**

20 A particle A of mass 2*m*, moving with speed 2*u* in a straight line on a smooth horizontal table, collides with a particle B of mass 3*m*, moving with speed *u* in the same direction as A. The coefficient of restitution between A and B is *e*.

a Show that the speed of B after the collision is

$\frac{1}{5}u(7 + 2e)$.

b Find the speed of A after the collision, in terms of *u* and *e*.

The speed of A after the collision is $\frac{11}{10}u$.

c Show that $e = \frac{1}{2}$.

At the instant of collision, A and B are at a distance *d* from a vertical barrier fixed to the surface at right-angles to their direction of motion. Given that B hits the barrier, and that the coefficient of restitution between B and the barrier is $\frac{11}{16}$,

d find the distance of A from the barrier at the instant that B hits the barrier,

e show that, after B rebounds from the barrier, it collides with A again at a distance $\frac{5}{32}d$ from the barrier.   **E**

21

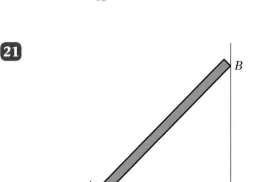

The diagram shows a uniform heavy plank of wood AB, of mass m, whose lower end A is resting on rough horizontal ground and whose upper end B is resting against a rough vertical wall. The coefficient of friction between the plank and the ground and between the plank and the wall is $\frac{2}{3}$. The plank is about to slip at both ends.

a Suggest a suitable model for the plank so that the forces exerted on it by the ground and the wall can be found.

b Show that the horizontal component of the force exerted by the wall on the plank is $\dfrac{6mg}{13}$.   **E**

22 A uniform rod, of mass m, rests with one end A against a rough vertical wall and the other end B on a rough horizontal floor. The vertical plane through the rod is perpendicular to the wall. The coefficient of friction between the wall and the rod is μ_1. The coefficient of friction between the floor and the rod is μ_2. Given that θ is the inclination of the rod to the floor when the rod is on the point of slipping, show that

$$2\mu_2 \tan \theta = 1 - \mu_1\mu_2.$$ **E**

23

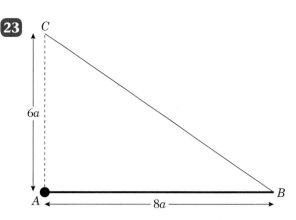

A uniform rod AB, of length 8a and weight W, is free to rotate in a vertical plane about a smooth pivot at A. One end of a light inextensible string is attached to B. The other end is attached to point C which is vertically above A, with $AC = 6a$. The rod is in equilibrium with AB horizontal, as shown in the diagram.

a By taking moments about A, or otherwise, show that the tension in the string is $\frac{5}{6}W$.

b Calculate the magnitude of the horizontal component of the force exerted by the pivot on the rod.   **E**

24

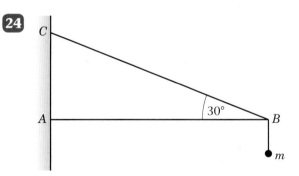

A uniform rod AB has mass 2m and length a. The end A is smoothly hinged at a fixed point. A particle of mass m is suspended from the rod at the end B. The loaded rod is held in equilibrium in a horizontal position by a light string, one end of which is attached to the rod at B, the other end being fixed to a point C vertically above A, as shown in the diagram. The string makes an angle of 30° with the horizontal.

a Show that the tension in the string is 4*mg*.

b Find the magnitude of the force exerted by the hinge on the rod at *A*. **E**

25

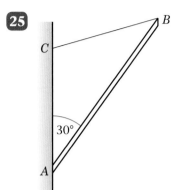

A uniform rod *AB* of mass *m* and length 2*a* is smoothly hinged to a vertical wall at *A* and is supported in equilibrium by a rope which is modelled as a light string. One end of the rope is attached to the end *B* of the rod and the other end is attached to a point *C* of the wall, where *C* is vertically above *A*, *AC* = *CB*, and ∠*CAB* = 30°, as shown in the diagram.

a Show that the tension in the rope is 0.5*mg*.

b Find the magnitude of the vertical component of the force acting on the rod at *A*.

c If the rope were not modelled as a light string, state how this would affect the tension throughout the rope. **E**

26 A uniform ladder *AB*, of mass *m* and length 2*a*, has one end *A* on rough horizontal ground. The coefficient of friction between the ladder and the ground is 0.6. The other end *B* of the ladder rests against a smooth vertical wall.

A builder of mass 10*m* stands at the top of the ladder. To prevent the ladder from slipping, the builder's friend pushes the bottom of the ladder horizontally towards the wall with a force of magnitude *P*. This

force acts in a direction perpendicular to the wall. The ladder rests in equilibrium in a vertical plane perpendicular to the wall and makes an angle α with the horizontal, where $\tan \alpha = \frac{3}{2}$.

a Show that the reaction of the wall on the ladder has magnitude 7*mg*.

b Find, in terms of *m* and *g*, the range of values of *P* for which the ladder remains in equilibrium. **E**

27

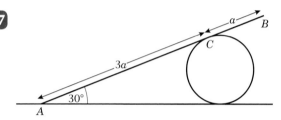

A piece of equipment used in an acrobatic show consists of a smooth cylinder which is fixed, with its axis horizontal, to a rough horizontal plane. A plank, which is modelled as a uniform rod *AB* of mass *m* and length 4*a*, rests in equilibrium on the cylinder at the point *C*, where *AC* = 3*a*. The end *A* of the plank rests on the plane and *AB* makes an angle of 30° with the horizontal, as shown in the diagram. The points *A*, *B* and *C* lie in a vertical plane which is perpendicular to the axis of the cylinder.

a Find the magnitude of the force exerted on the plank by the cylinder at the point *C*.

Given that the plank is in limiting equilibrium and that the coefficient of friction between the plank and the plane is μ,

b show that $\mu = \frac{1}{3}\sqrt{3}$. **E**

28 A non-uniform ladder *AB*, of length 4*a* and weight *W*, has its centre of mass at a distance *a* from *B*. The ladder rests with *A* against a rough vertical wall and with its lower end *B* on smooth horizontal ground. The coefficient of friction between the wall and the ladder is μ. The ladder is in a vertical plane perpendicular to the wall and makes an angle α with the horizontal where $\tan \alpha = 2$. A man can just prevent the ladder from slipping down by applying a horizontal force of magnitude P, perpendicular to the wall, at B. The ladder is modelled as a non-uniform rod.

a Draw a diagram showing all the forces acting on the ladder.

b Find an expression for P in terms of W and μ. **E**

29

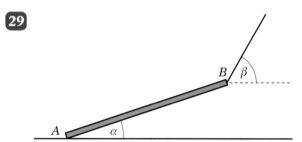

A straight log AB has weight W and length $2a$. A cable is attached to one end B of the log. The cable lifts the end B off the ground. The end A remains in contact with the ground, which is rough and horizontal. The log is in limiting equilibrium. The log makes an angle α to the horizontal, where $\tan \alpha = \frac{5}{12}$. The cable makes an angle β to the horizontal, as shown in the diagram. The coefficient of friction between the log and the ground is 0.6. The log is modelled as a uniform rod and the cable as light.

a Show that the normal reaction on the log at A is $\frac{2}{5}W$.

b Find the value of β.

The tension in the cable is kW.

c Find the value of k. **E**

30

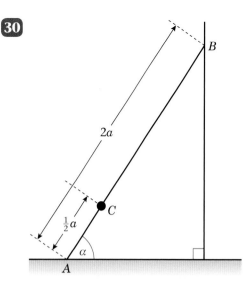

A uniform ladder AB, of mass m and length $2a$, has one end A on rough horizontal ground. The other end B rests against a smooth vertical wall. The ladder is in a vertical plane perpendicular to the wall. The ladder makes an angle α with the horizontal, where $\tan \alpha = \frac{4}{3}$. A child of mass $2m$ stands on the ladder at C where $AC = \frac{1}{2}a$, as shown in the diagram. The ladder and the child are in equilibrium.

By modelling the ladder as a rod and the child as a particle, calculate the least possible value of the coefficient of friction between the ladder and the ground. **E**

31

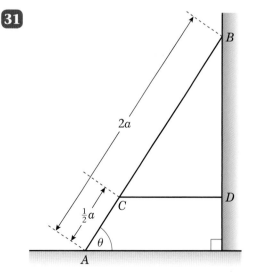

A uniform ladder, of weight W and length $2a$, rests in equilibrium with one end A on a smooth horizontal floor and the other end B on a rough vertical wall. The ladder is in a vertical plane perpendicular to the wall. The coefficient of friction between the wall and the ladder is μ. The ladder makes an angle θ with the floor, where $\tan\theta = 2$. A horizontal light inextensible string CD is attached to the ladder at the point C, where $AC = \frac{1}{2}a$. The string is attached to the wall at the point D, with BD vertical, as shown in the diagram. The tension in the string is $\frac{1}{4}W$. By modelling the ladder as a rod,

a find the magnitude of the force of the floor on the ladder,

b show that $\mu \geqslant \frac{1}{2}$.

c State how you have used the modelling assumption that the ladder is a rod. **E**

32 A uniform ladder AB, of mass 10 kg and length 4 m, rests in equilibrium with the end A on rough horizontal ground. The end B of the ladder rests against a smooth vertical wall, the ladder being in a vertical plane perpendicular to the wall. The coefficient of friction between the ladder and the ground is $\frac{1}{3}$. The ladder is inclined at an angle θ to the horizontal, where $\tan\theta = 2$. A man of mass 80 g stands on the ladder at a point which is a distance x metres from A.

Find the range of possible values of x. **E**

33 A uniform ladder, of mass M and length 5 m, has one end on rough horizontal ground with the other end placed against a smooth vertical wall. The coefficient of friction between the ladder and the ground is 0.3. The highest point of the wall is higher than the highest point on the ladder. Given that the top of the ladder is 4 m vertically above the level of the ground,

a show that the ladder cannot remain in equilibrium in this position.

A brick is placed on the bottom rung of the ladder in order to enable it to stay in equilibrium in the position described above. Assuming that the brick is at the very bottom of the ladder and does not touch the ground,

b show that the horizontal frictional force exerted on the ladder by the ground is independent of the mass of the brick.

c Find, in terms of M, the smallest mass of the brick which will enable the ladder to remain in equilibrium.

The ladder, without the brick, is now extended so that the top of the ladder is higher than the top of the wall.

d Draw a diagram showing the forces acting on the ladder in this situation. **E**

34

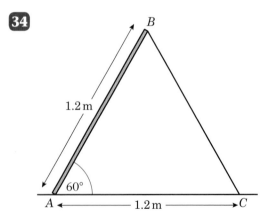

A trap door is propped open at 60° to the horizontal by a pole. The trap door is modelled as a uniform rod AB, of mass 10 kg and length 1.2 m, smoothly hinged at A. The pole is modelled as a light rod BC, smoothly hinged to AB at B. The points A and C are at the same horizontal level, $AC = 1.2$ m and the plane ABC is vertical, as shown in the diagram.

Find, to 3 significant figures,

a the thrust in BC,

b the magnitude of the force acting on the rod AB at A.   **E**

35

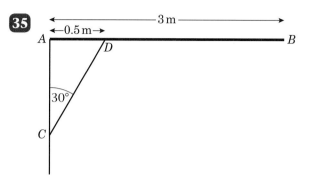

A uniform pole AB, of mass 30 kg and length 3 m, is smoothly hinged to a vertical wall at one end A. The pole is held in equilibrium in a horizontal position by a light rod CD. One end C of the rod is fixed to the wall vertically below A. The other end D is freely jointed to the pole so that $\angle ACD = 30°$ and $AD = 0.5$ m, as shown in the diagram. Find

a the thrust in the rod CD,

b the magnitude of the force exerted by the wall on the pole at A.

The rod CD is removed and replaced by a longer light rod CM, where M is the mid-point of AB. The rod is freely jointed to the pole at M. The pole AB remains in equilibrium in a horizontal position.

c Show that the force exerted by the wall on the pole at A now acts horizontally.   **E**

36 A uniform ladder rests with its lower end on a rough horizontal path and its upper end against a smooth vertical wall. The ladder rests in a vertical plane perpendicular to the wall. A woman stands on the top of this ladder, and the ladder is in limiting equilibrium. The weight of the woman is twice the weight of the ladder, and the coefficient of friction between the

path and the ladder is $\frac{5}{12}$. By modelling the ladder as a uniform rod and the woman as a particle, find, to the nearest degree, the angle between the ladder and the horizontal.   **E**

37 A uniform rod AB, of length $2a$ and weight W, is hinged to a vertical post at A and is supported in a horizontal position by a string attached to B and to a point C vertically above A, where $\angle ABC = \theta$. A load of weight $2W$ is hung from B. Find the tension in the string and the horizontal and vertical resolved parts of the force exerted by the hinge on the rod. Show that, if the reaction of the hinge at A is at right angles to BC, then $AC = 2a\sqrt{5}$.   **E**

38

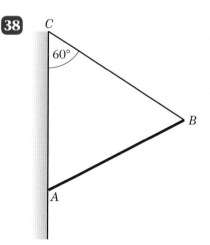

A uniform rod AB of mass m rests in equilibrium with A in contact with a rough vertical wall. The coefficient of friction between the rod and the wall is μ. A light string is attached to B and to a point C of the wall, where C is vertically above A. The plane ABC is perpendicular to the wall, $BC = BA$ and $\angle ACB = 60°$, as shown in the diagram.

a Show that the tension in the string is $\frac{1}{2}mg$.

b Show that $\mu \geqslant \sqrt{3}$.   **E**

39

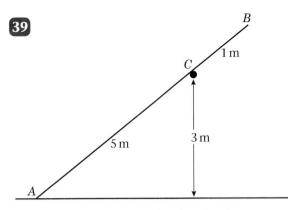

A smooth horizontal rail is fixed at a height of 3 m above a horizontal playground whose surface is rough. A straight uniform pole AB, of mass 20 kg and length 6 m, is placed to rest at a point C on the rail with the end A on the playground. The vertical plane containing the pole is at right angles to the rail. The distance AC is 5 m and the pole rests in limiting equilibrium as shown in the diagram. Calculate

a the magnitude of the force exerted by the rail on the pole, giving your answer to the nearest N,

b the coefficient of friction between the pole and the playground, giving your answer to 2 decimal places,

c the magnitude of the force exerted by the playground on the pole giving your answer to the nearest N. **E**

40 A pole of mass m and length $2a$ is used to display a light banner. The pole is modelled as a uniform rod AB, freely hinged to a vertical wall at the point A. It is held in a horizontal position by a light wire. One end of the wire is attached to the end B of the rod and the other end is attached to the wall at a point C which is vertically above A such that $\angle ABC$ is θ, where $\tan \theta = \frac{1}{2}$, as shown in the diagram.

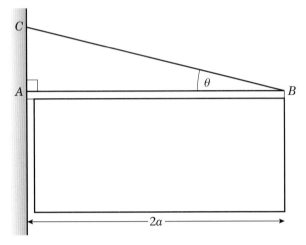

a Show that the tension in the wire is $\dfrac{mg}{2 \sin \theta}$.

b Find, in terms of m and g, the magnitude of the force exerted by the wall on the rod at A.

c State, briefly, where in your calculation you have used the modelling assumption that the pole is a rod. **E**

Practice exam paper

Whenever a numerical value of g is required, take $g = 9.8\,\mathrm{m\,s^{-2}}$.

1 A uniform rod AB has mass $2\,\mathrm{kg}$ and length $0.8\,\mathrm{m}$. The end A of the rod is hinged to a fixed point. A light inextensible string has one end attached to B and lies in the same vertical plane as the rod. The string is perpendicular to the rod. The rod is in equilibrium, inclined at an angle α to the horizontal, where $\tan \alpha = \frac{3}{4}$, with B higher than A, as shown in the figure.

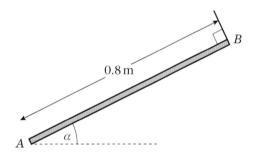

Find

a the tension in the string, (3)

b the magnitude and direction of the force acting on the rod at A. (8)

2 A box of mass $2\,\mathrm{kg}$ is projected down a rough inclined plane with speed $4\,\mathrm{m\,s^{-1}}$. The plane is inclined to the horizontal at an angle β, where $\tan \beta = \frac{4}{3}$, and the coefficient of friction between the box and the plane is 0.5. The box is modelled as a particle and air resistance is negligible. Using the work–energy principle, find the speed of the box after it has moved a distance of $5\,\mathrm{m}$ down the plane. (7)

3 A uniform lamina $ABCD$ is in the shape of a trapezium with $AB = 4a$, $CD = a$, $AD = 2a$ and AD perpendicular to both AB and DC, as shown in the figure.

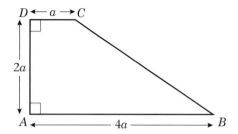

a Find the distance of the centre of mass of the lamina from

   i AD **ii** AB.   (6)

The lamina is freely suspended by a string attached to the point A and hangs at rest.

b Find, to the nearest degree, the angle between CD and the vertical.   (4)

4 A man throws a ball from the top of a vertical cliff which is 30 m high. The ball rises to a height of 10 m above the level of the top of the cliff before falling and it strikes the sea at a distance of 90 m from the foot of the cliff as shown in the figure.

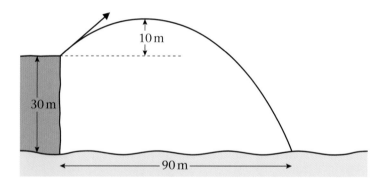

The motion of the ball is modelled as that of a particle moving freely under gravity. Find

a the vertical component of the initial velocity of the ball,   (3)

b the time of flight,   (4)

c the horizontal component of velocity,   (2)

d the direction of motion of the ball as it strikes the sea.   (5)

5 A particle P of mass m is at rest on a smooth horizontal plane at a distance of 2 metres from a smooth vertical wall. Another particle Q, also of mass m, is moving with speed u along the plane in a straight line towards the wall. The direction of motion of Q is perpendicular to the wall. Particle Q strikes particle P directly. The coefficient of restitution between P and Q is $\frac{1}{3}$.

  a Show that, after P and Q have collided,

    i the speed of P is $\frac{2u}{3}$,

    ii the speed of Q is $\frac{u}{3}$.   (7)

  Particle P then goes on to strike the wall. The coefficient of restitution between P and the wall is also $\frac{1}{3}$.

  b Show that the second collision between P and Q occurs at a distance of 0.4 m from the wall.   (7)

6 A particle P moves in a straight line in such a way that, at time t seconds, its acceleration $a\,\mathrm{m\,s}^{-2}$ is given by

    $a = 6t - 3t^2, t \geqslant 0.$

  When $t = 0$, P is at rest at the point O.

  a Find the times when the particle is at rest.   (4)

  b Find the distance travelled by P during the first 5 seconds.   (5)

7 A car of total mass 1200 kg is travelling along a straight horizontal road at $40\,\mathrm{m\,s}^{-1}$, when the driver suddenly applies the brakes. The brakes exert a constant force and the car comes to rest after travelling a distance of 80 m. The other resistances on the car total 500 N.

  a Find the magnitude of the force exerted by the brakes on the car.   (4)

  A trailer, with no brakes, is now attached to the car by means of a tow-bar. The mass of the trailer is 600 kg, and when the trailer is moving, it experiences a constant resistance of magnitude 420 N. The tow-bar is modelled as a light rod which remains parallel to the road at all times. The car and trailer come to a straight hill, which is inclined to the horizontal at an angle α, where $\sin \alpha = \frac{1}{14}$. They move together down the hill. The driver suddenly applies the brakes, which exert a force of the same magnitude as before. The other resistances on the car remain at 500 N.

  b Find the deceleration of the car and the trailer when the brakes are applied.   (3)

  c Find the magnitude of the force exerted on the car by the trailer when the brakes are applied.   (3)

Answers

Exercise 1A

1 3.1 (2 s.f.)
2 8.5 m (2 s.f.)
3 **a** 31 m (2 s.f.)   **b** 80 m
4 **a** 2.7 s (2 s.f.)   **b** 790 m (2 s.f.)
5 **a** 10 m (2 s.f.)   **b** 41 m (2 s.f.)
6 **a** 3.9 s (2 s.f.)   **b** 56 m (2 s.f.)
7 55° (nearest degree)
8 77 m s^{-1} (2 s.f.)
9 **a** $(36\mathbf{i} + 27.9\mathbf{j})$ m   **b** 13 m s^{-1} (2 s.f.)
10 **a** 22° (2 s.f.)   **b** 97 m (2 s.f.)
11 **a** 16 (2 s.f.)   **b** 1.6 s (2 s.f.)
12 **a** 4.4   **b** 88   **c** 50° (2 s.f.)
13 **a** 1.1 s (2 s.f.)
    b 34 m (2 s.f.)

14 **b** $\tan \alpha = \frac{5}{4}$

15 **d** 12° and 78° (nearest degree)
16 $\alpha = 40.6°$ (nearest 0.1°)
    $u = 44$ (2 s.f.)

Exercise 1B

1 **a** 46 m s^{-1} **b** 24 m s^{-2}
2 **a** 16 m s^{-2} **b** 12 m
3 **a** $v = 6 + 16t - t^2$ **b** -6
4 7 m s^{-2} in the direction of x decreasing.
5 6.75 m
6 **a** 17 m s^{-2} **b** 12.375 m
7 **a** $10\frac{2}{3}$ **b** 13 m
8 **a** $t = \frac{3}{2}, 5$
    b $t = \frac{3}{2}$, 8 m s^{-2} in the direction of x decreasing
      $t = 5$, 34 m s^{-2} in the direction of x increasing
9 **a** $k = -4$ **b** 4 m   **c** 0.05
10 **a** 7 m s^{-1} in the direction of x decreasing
    b $t = 1, 5$
    c 6 m

11 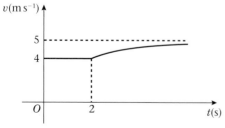

    b 21.8 m
12 **a** $\frac{28}{3}$ m s^{-1} **b** $\frac{58}{3}$ m s^{-1} **c** 36 m

Exercise 1C

1 **a** $(3\mathbf{i} + 23\mathbf{j})$ m s^{-1} **b** $18\mathbf{j}$ m s^{-2}
2 **a** $(2t\mathbf{i} + 2\mathbf{j})$ m s^{-2} **b** $(3\frac{1}{3}\mathbf{i} + 2\mathbf{j})$ m
3 **a** $16\mathbf{i}$ m s^{-1} **b** 140.8 m
4 **a** 20 m s^{-1} **b** 10 m s^{-2}
5 **a** $6\sqrt{5}$ m s^{-1} **b** $t = 2$ **c** $(-16\mathbf{i} + 4\mathbf{j})$ m
6 **a** $((3t^2 + 9)\mathbf{i} + (4t^3 - 10t + 6)\mathbf{j})$ m s^{-1}
    b 34° (nearest degree)
7 **a** 13 m   **b** $(4\mathbf{i} + 6\mathbf{j})$ m s^{-2}
8 **a** $\mathbf{v} = (t^2 - 4t)\mathbf{i} + (6t - 24)\mathbf{j}$
    b 7.81 m s^{-1} (3 s.f.)
9 **a** $k = -0.5, -8.5$
    b 10 m s^{-2} for both values of k
10 **a** $\left(\left(\frac{5t^2}{2} - 3t + 2\right)\mathbf{i} + \left(8t - \frac{t^2}{2} - 5\right)\mathbf{j}\right)$ m s^{-1}
    b $t = \frac{1}{2}$ **c** $\frac{9\sqrt{2}}{8}$ m s^{-1}
11 **a** $\left((t^2 + 6)\mathbf{i} + \left(2t - \frac{t^3}{3}\right)\mathbf{j}\right)$ m   **b** $6\mathbf{j}$ m
12 **a** $(45\mathbf{i} + 67.5\mathbf{j})$ m   **b** $(135\mathbf{i} - 90\mathbf{j})$ m s^{-1}

Mixed exercise 1D

1 **a** 45 m   **b** 6.1 s (2 s.f.)
2 $h = 35$ (2 s.f.)
3 **a** $t = 5$ **b** 37.5 m
4 **a** 30 m s^{-2} **b** $OP = 75$ m
5 **b** 10 m s^{-2}
      143.1° (nearest 0.1°)
6 **a** $(-\mathbf{i} - \frac{1}{4}\mathbf{j})$ m s^{-1} **b** $(94\mathbf{i} - 654\mathbf{j})$ m
7 **a** 36 m   **b** 30 m (2 s.f.)
8 **a** 140 (2 s.f.)   **b** 36 m s^{-1} (2 s.f.)
9 **a** $-6\mathbf{i}$ m s^{-1}
10 0.8 N
11 **a** $((2t^4 - 3t^2 - 4)\mathbf{i} + (4t^2 - 3t - 7)\mathbf{j})$ m s^{-1}
    b $t = \frac{7}{4}$
12 **b** 4, $\frac{234}{29} \approx 8.07$
13 **a** $0.3\sqrt{3}$ m   **b** $t = 3$ **c** 0.329 m s^{-2} (3 s.f.)
14 **a** $(\ln 2 - 2)$ m s^{-2} in the direction of x increasing.
    b $\frac{8}{e}$ m
15 **b** 250 m
    c 22° (nearest degree)
16 **b** $\frac{u^2}{g}$
    c 20.9°, 69.1° (nearest 0.1°)
17 **a** 2.0 s (2 s.f.)   **b** 3.1 s (2 s.f.)
    c 36 m s^{-1} (2 s.f.)

18 a $(-3t^2 + 22t - 24)\,\text{m s}^{-1}$ **b** $t = \frac{4}{3}, 6$ **c** $\frac{11}{3}$

d

e $0 \leqslant t < 0.38, \frac{10}{3} < t < 4$

19 a $24\,\text{m s}^{-1}$ in the direction of s increasing
b $24\,\text{m s}^{-1}$ in the direction of s decreasing
c $132\,\text{m}$
d 20.25 and 118.5

20 a $t = 2$ **b** $(24\mathbf{i} - 8\mathbf{j})\,\text{m}$

Exercise 2A

1 $(3.2, 0)$ **2** $(0, 2.5)$
3 $(1.1, 0)$ **4** $2\frac{1}{3}\,\text{m}$
5 $m = 6$ **6** $0.7\,\text{kg}$
7 $6\frac{1}{2}$ **8** $(0, -2)$
9 $m_1 = 2$ **10** $(4.8, 6.4)$
$m_2 = 3$

Exercise 2B

1 $(3, 2)$
2 $(\frac{1}{2}, -\frac{3}{4})$
3 $(4.6, 4.2)$
4 $(3\mathbf{i} + 2.5\mathbf{j})$
5 $(2.1, 0.3)$
6 a 1 **b** 1.5
7 $p = 1$ $q = -2$
8 $(1, 3)$
9 a $2\frac{2}{3}\,\text{cm}$ **b** $3\frac{1}{2}\,\text{cm}$
10 a $3g$ **b** $3.2\,\text{cm}$

Exercise 2C

1 a $(2, 3)$ **b** $(3, 4)$ **c** $(\frac{1}{3}, 1)$ **d** $\left(\frac{8a}{3}, 3a\right)$

2 Centre of mass is on the axis of symmetry at a distance $\frac{16}{3\pi}\,\text{cm}$ from the centre.

3 $a = 3$ $b = 3$

4 a Distance a from AB Distance $\frac{2a}{3}$ from BC.

b Distance $\frac{a}{3}$ from BC Distance $\frac{4a}{3}$ from AB.

c On line of symmetry, $\frac{4a}{3}$ from the line AB

d $\left(3a, \frac{2a}{3}\right)$ with A as origin and AC as x-axis

e $\left(\frac{8a}{3}, a\right)$ with A as origin and AC as x-axis

f $\left(2a, \frac{2a\sqrt{3}}{3}\right)$ with A as origin and AC as x-axis

5 $(2, 7)$ and $(-13, 4)$

Exercise 2D

1 $(2\frac{1}{2}, \frac{13}{14})$ **2** $(\frac{11}{4}, \frac{5}{4})$
3 $(1.7, 2.6)$ **4** $(\frac{79}{26}, \frac{51}{26})$
5 $(\frac{113}{30}, \frac{49}{30})$ **6** $(\frac{32}{13}, \frac{31}{13})$
7 $(\frac{7}{3}, 2)$ **8** $(3, 3)$
9 $(\frac{32}{9}, \frac{53}{18})$ **10** $(\frac{107}{27}, \frac{94}{27})$
11 $(2, \frac{63}{17})$ **12** $(\frac{319}{81}, \frac{286}{81})$
13 $(\frac{25}{8}, 3)$ **14** $\left(\frac{27\pi - 7}{9\pi - 2}, \frac{9\pi + 31}{9\pi - 2}\right)$
15 $\left(\frac{18 + 9\pi}{4 + 3\pi}, 3\right)$ **16** $\left(\frac{36 - \pi}{18 - \pi}, \frac{36 - 2\pi}{18 - \pi}\right)$

Exercise 2E

1 $(2\frac{1}{2}, 1)$ **2** $(\frac{43}{16}, \frac{21}{16})$
3 $(\frac{13}{8}, \frac{19}{8})$ **4** $(\frac{53}{18}, \frac{20}{9})$
5 $(\frac{18}{5}, \frac{9}{5})$
6 Centre of mass is on line of symmetry through O, and a distance of $\dfrac{9(\sqrt{3} + 2)}{6 + \pi}$ from O.

7 Centre of mass is on the line of symmetry at a distance of $\dfrac{3}{2\pi}$ below the line AB.

Exercise 2F

1 a $20.4°$ (3 s.f.) **b** $24.4°$ (3 s.f.)
c $56.8°$ (3 s.f.)
2 $63.0°$ (3 s.f.) **3** $80.5°$ (3 s.f.)
4 $33.1°$ (3 s.f.) **5** $81.0°$ (3 s.f.)
6 $\alpha = 53°$
7 a i $\frac{26}{7}\,\text{cm}$ **ii** $\frac{18}{7}\,\text{cm}$
b $\alpha = 34.7°$

Mixed exercise 2G

1 a $0.413\,\text{m}$ (3 s.f.) **b** $12°$ (nearest degree)
c 0.275 (3 s.f.)
2 $\theta = 36.9°$
3 $(-\frac{1}{7}, \frac{3}{2})$
4 a $\dfrac{13a}{9}$ **b** $\dfrac{4a}{9}$
c $45°$ **d** $m = \dfrac{5M}{9}$
5 a i $\dfrac{4a}{5}$ **ii** $\dfrac{a}{2}$ **b** $\theta = 58°$
6 a i $1.7a$ **ii** $1.1a$ **b** $\theta = 32.5°$

Exercise 3A

1 $2.52\,\text{J}$ **2** $8.5\,\text{N}$
3 $24.0\,\text{J}$ **4** $588\,\text{J}$
5 $330\,\text{J}$ **6** $73.5\,\text{J}$
7 $38.3\,\text{m}$ **8** $228\,\text{J}$
9 0.255 **10** $64.7\,\text{J}$
11 $23\,400\,\text{J}$ **12** $281\,\text{J}$
13 $2.48\,\text{J}$
14 a $21.7\,\text{N}$ **b** $326\,\text{J}$ **c** $452\,\text{J}$
15 $16.5\,\text{J}$
16 0.559
17 $112\,\text{J}$

Exercise 3B

1 a 33.8 J   b 6 J   c 141 J
   d 10 000 J   e 200 J   f 45 J
   g 160 000 J
2 a 44.1 J   b 8085 J   c 22 050 J
   d 34 104 J   e 24 696 J   f 70.6 J
   g 125 440 J
3 76.8 J   4 168 750 J
5 8   6 $4.53\,\mathrm{m\,s^{-1}}$
7 728 J
8 a 11.8 J   b 4.85 J
9 384 000 J
10 a 65 625 J   b 1 837 500 J
11 20.0 m

Exercise 3C

1 a 27.4 J   b $11.7\,\mathrm{m\,s^{-1}}$
2 a 36 J   b 36 J   c 7.35 m
3 a 56.3 J   b 56.3 J   c 5.63 m
4 a 9.6 J   b 9.6 J   c 0.350
5 a 54 J   b 54 J   c 4.59 m
6 $9.90\,\mathrm{m\,s^{-1}}$ 7 20.4 m
8 10.6   9 250 000 N (or 250 kN)
10 0.075 m (or 75 mm)
11 a 56.2 J   b 56.2 J   c $4.74\,\mathrm{m\,s^{-1}}$
12 0.408   13 8.27
14 $7.94\,\mathrm{m\,s^{-1}}$ 15 2.33 m
16 a 5400 J   b 13.8 m
17 128 N   18 11.8 m

Exercise 3D

1 18 kW   2 15 000 W (or 15 kW)
3 278 N   4 $25\,\mathrm{m\,s^{-1}}$
5 20 000 W (or 20 kW)   6 550 N
7 a $1.10\,\mathrm{m\,s^{-2}}$ b $0.294\,\mathrm{m\,s^{-2}}$ c $25.7\,\mathrm{m\,s^{-1}}$
8 11 400 W (or 11.4 kW)   9 $R = 300$
10 $10\,\mathrm{m\,s^{-1}}$
11 a 175 N   b $0.854\,\mathrm{m\,s^{-2}}$
12 a $0.868\,\mathrm{m\,s^{-2}}$ b $13.2\,\mathrm{m\,s^{-1}}$
13 a 18 000 W (or 18 kW)   b $6.80\,\mathrm{m\,s^{-1}}$
14 37.9
15 a $6.11\,\mathrm{m\,s^{-1}}$ b $0.342\,\mathrm{m\,s^{-2}}$

Mixed exercise 3E

1 20.2 N
2 a 2940 J   b $98\,\mathrm{J\,s^{-1}}$ (or 98 W)
3 a 20 J   b 0.163
4 a $4.48\,\mathrm{m\,s^{-2}}$ b 1.51 m
5 a $0.708\,\mathrm{m\,s^{-2}}$ b $0.521\,\mathrm{m\,s^{-2}}$
6 a 11.4 kW   b 21.3
7 a $\dfrac{9mgs}{5}$ b $\dfrac{14gs}{25}$
8 a $2.95\,\mathrm{m\,s^{-1}}$ b 61.2 J
9 32 600 000 J (or 32 600 kJ)
10 a 16.2 J   b 16.2 J   c 4.05 N
11 a 250 J   b 0.638
12 a 480 N   b $25.4\,\mathrm{m\,s^{-1}}$
13 a 7.42 N   b 435 J   c $14.0\,\mathrm{m\,s^{-1}}$
14 a $33.3\,\mathrm{m\,s^{-1}}$ b $0.222\,\mathrm{m\,s^{-2}}$
15 a 12 kW   b 24 kW   c 10.8

Review Exercise 1

1 a 73 m (2 s.f.)   b 87 m (2 s.f.)
   c $43\,\mathrm{m\,s^{-1}}$ (2 s.f.)
     66° (nearest degree), below the horizontal
2 a 1.8 m   b $6.8\,\mathrm{m\,s^{-1}}$ (2 s.f.)
3 a 48 m   b 120 m (2 s.f.)
   c $T = 2.5$ s, $\mathbf{r} = (20\mathbf{i} - \frac{45}{8}\mathbf{j})$ m
4 b $u = 7$ c 1.75
5 b 3.5 s
6 a 1.05 seconds (2 d.p.)
   c 12 (2 s.f.)
   d B and T are not particles but take up space; they have extension. This would allow a range of values of V resulting in hitting the target.
7 b $OA = 500$ m
    The greatest height reached is 125 m.
   c $70\,\mathrm{m\,s^{-1}}$
   d 10 s (nearest second)
8 b 10° and 78°
   c 2.0 s (2 s.f.)
9 6 s
10 a 25 m   b 12.5 m
11 a $k = \frac{1}{2}$ b $t = \pi, 3\pi$
12 a $45.6\,\mathrm{m\,s^{-1}}$ 151.2 m
   b 396.9 m
   c $\frac{2}{15}T^3 - 5T^2 + 396.9 = 0$
   d The aircraft satisfies the safety condition.
13 b The greatest speed is $40\frac{1}{3}\,\mathrm{m\,s^{-1}}$.
   c 196 m
14 b $\frac{441}{12}$ m
15 a 4.02 (2 d.p.)   b $(67\mathbf{i} + 28\mathbf{j})$ m
16 a $52\,\mathrm{m\,s^{-1}}$ b $(12\mathbf{i} + \frac{15}{2}\mathbf{j})\,\mathrm{m\,s^{-2}}$
17 a 4   b $(-36\mathbf{i} + 8\mathbf{j})\,\mathrm{m\,s^{-2}}$
18 a 5.01 (2 d.p.)   b 78.45 m (2 d.p.)
19 a P is $(6t\mathbf{i} + 2\mathbf{j})\,\mathrm{m\,s^{-1}}$
    Q is $(\mathbf{i} + 3t\mathbf{j})\,\mathrm{m\,s^{-1}}$
   b $12.2\,\mathrm{m\,s^{-1}}$ (3 s.f.)
   c $t = \frac{1}{3}$
   d $(7\mathbf{i} + \frac{3}{2}\mathbf{j})$ m
20 c $-6\omega^2 \sin \omega t\mathbf{i} - 4\omega^2 \cos \omega t\mathbf{j}$
   d 109.8° (1 d.p.)
21 a $\lambda = 2$ b $k = -1.1$
22 $x = y = \frac{1}{2}$
23 $(3\mathbf{i} + 2.5\mathbf{j})$ m
24 b $c = -\frac{1}{3}$
25 a 10.7 cm (3 s.f.)   b 25° (nearest degree)
26 a 3   b 2   c 37° (nearest degree)
27 a i $\frac{5l}{12}$ ii $\frac{l}{3}$ b 39° (nearest degree)
28 a 3   b $\frac{1}{7}$
29 b $\frac{11}{15}$
30 a $\frac{6l}{5}$ b l c 51° (nearest degree)
31 a $\frac{19}{15}a$ b $\frac{7}{45}M$
32 a i $\frac{5}{2}a$ ii $\frac{4}{3}a$ b 15° (nearest degree)
33 a i $\frac{5}{4}d$ ii $\frac{3}{4}d$ b 23° (nearest degree)
34 a 6 cm   b 22.6° (1 d.p.)
35 a 6.86 cm (2 d.p.)   b 32.1° (1 d.p.)
36 a From AD, $\frac{7}{9}a$, from AB, $\frac{5}{9}a$
   b $\frac{2}{5}$

37 **a** 18° (nearest degree) **b** $\frac{1}{4}M$

38 **a** 25 cm **b** $\frac{3}{11}m$

39 **a** $\left(\frac{9}{2}, \frac{2}{3}\right)$

  c $\frac{4}{9}$ **d** 83.7° (1 d.p.)

40 **b** 0.7*a* **c** 20° (nearest degree)

  e $\frac{5\sqrt{65}}{4}mg$

41 **a** 2*u* **b** 15*mu*²

42 6.25

43 **a** 8.4 m s⁻¹ (2 s.f.) **b** 0.42 (2 s.f.)

44 **a** 41 J (2 s.f.) **b** 0.67 (2 s.f.)

45 **a** 22.4 J **b** 6.4 m s⁻¹ (2 s.f.)

  c 4.3 m s⁻¹ (2 s.f.)

46 **a** $\frac{7}{5}mgh$

  b $\frac{3}{5}gh$

47 400 N

48 87 kW (2 s.f.)

49 220

50 **a** 0.7 m s⁻² **b** 44.4 kW

51 **a** 35 m s⁻¹ **b** 15 m s⁻¹ (2 s.f.)

52 **a** 960 N **b** 0.75 **c** 360 N

53 **b** 20 (2 s.f.)

54 **a** 0.8 **b** $81\frac{2}{3}$

  c Resistance usually varies with speed.

55 **a** 2200 J (2 s.f.) **b** 300 W

56 **a** 0.15 m s⁻² **b** 36 m s⁻¹ (2 s.f.)

  c The resistance usually increases with speed.

57 **a** $P = 8(R + 20)$ **b** 480 **c** 240

58 **b** 1.4 m s⁻²

  c 850 N

  d 335 kJ (3 s.f.)

  e The resistance could be modelled as varying with speed.

59 **b** $\psi = 55°$ (nearest degree)

  c 60 m

60 **a** 52 (2 s.f.) **b** 3 s

  c 48 m **d** 24 m s⁻¹ (2 s.f.)

Exercise 4A

1 $(44\mathbf{i} - 24\mathbf{j})$ m s⁻¹ **2** $(8\mathbf{i} + 8\mathbf{j})$ m s⁻¹
3 $(\mathbf{i} - 2\mathbf{j})$ m s⁻¹ **4** $(3\mathbf{i} - 4\mathbf{j})$ m s⁻¹
5 $5(18\mathbf{i} - 24\mathbf{j})$ N s, $(7\mathbf{i} - 7\mathbf{j})$ m s⁻¹
6 $(10\mathbf{i} - 5\mathbf{j})$ N s, $(25\mathbf{i} + 2\mathbf{j})$ m s⁻¹
7 $(-12\mathbf{i} - 12\mathbf{j})$ N s **8** $(-6\mathbf{i} + 4\frac{1}{2}\mathbf{j})$ N s
9 $|\mathbf{Q}| = 30$ $\alpha = 37°$ (nearest degree)
10 $|\mathbf{Q}| = \sqrt{5} = 2.24$ (3 s.f.) $\alpha = 27°$ (nearest degree)
11 $6\sqrt{10}$ or 19.0 N s (3 s.f.)
12 $(-5\mathbf{i} + 30\mathbf{j})$ m s⁻¹
13 $\mathbf{v} = (14\mathbf{i} + 20\mathbf{j})$ m s⁻¹
14 $\mathbf{v} = \mathbf{i} + 2\mathbf{j}$ 18° (nearest degree)
15 $6\mathbf{i}$ m s⁻¹
16 0.606 m s⁻¹ (3 s.f.),   45°

Exercise 4B

1 **a** $\frac{2}{3}$ **b** $\frac{1}{2}$ **c** $\frac{1}{3}$
2 **a** $v_1 = 0, v_2 = 3$ **b** $v_1 = 2\frac{1}{2}, v_2 = 3$
  c $v_1 = 4, v_2 = 6$ **d** $v_1 = -4, v_2 = 4$
  e $v_1 = -5, v_2 = -2$
3 **a** 3.5 m s⁻¹ **b** 1

4 5 m s⁻¹ and 3 m s⁻¹ both in the direction that *B* was moving before the impact.
  18 N s

5 $\frac{u}{2}, \frac{1}{4}$

6 $\frac{u}{3}(5 - 4e), \frac{u}{3}(5 + 2e)$

8 **a** $\frac{u}{10}(10 - 3k)$

9 **a** $u(5 - 3k)$

10 **b** $\frac{u}{2}(5 - 3e)$

  d $e = \frac{1}{3}$

Exercise 4C

1 **a** $e = \frac{2}{5}$ **b** $e = \frac{1}{2}$
2 **a** 3.5 m s⁻¹ **b** 3 m s⁻¹
3 **a** 8 m s⁻¹ **b** 8 m s⁻¹
4 $e = 0.75$
5 0.77 (2 s.f.)
6 18.75 cm
7 $\frac{1}{2}$
8 2.94 s

Exercise 4D

1 **a** $u = 3, v = 5, x = 4, y = 4\frac{1}{2}$
  b $u = 2, v = 4, x = 3\frac{1}{2}, y = 4$
2 $-\mathbf{i}$ m s⁻¹, $0.5\mathbf{i}$ m s⁻¹ and $1.75\mathbf{i}$ m s⁻¹
3 -1.5 m s⁻¹, 0.5 m s⁻¹ and 5 m s⁻¹
4 **a** $\frac{1}{2}u(1 - e), \frac{1}{4}u(1 + e)(1 - e)$ and $\frac{1}{4}u(1 + e)^2$
  b *A* will catch up with *B* provided that $2 > 1 + e$
  Since $e < 1$ this condition holds and *A* will catch up with *B*, resulting in a further collision.
5 **a** $u(1 + 3e) > 3u \Rightarrow e > \frac{2}{3}$
  b *A* moves away from *B*.
6 $\frac{5u}{8}$ and $\frac{7u}{12}$
7 **a** 3*u* and 5*u*
  b $\frac{17u}{4}$ and $\frac{43u}{12}$
8 **a** 19.6 cm **b** 9.604 cm **c** 1.17 m (3 s.f.)
9 **a** e^2H **b** e^4H **c** $\frac{H(1 + e^2)}{(1 - e^2)}$

Exercise 4E

1 **a** 6 m s⁻¹ **b** 1.5 J
2 *A* has speed $\frac{7u}{3}$ away from *B* and *B* has speed $\frac{u}{3}$ towards *A*.
  Loss of K.E. is $\frac{5mu^2}{3}$
3 60 J
4 0.225 J
5 **a** 2 m s⁻¹ **b** 12 060 J = 12.06 kJ
6 $(8.5\mathbf{i} + \mathbf{j})$ m s⁻¹, 11.25 J
7 20.4 J
8 $(4\mathbf{i} + 20\mathbf{j})$ m s⁻¹   7.8 J
9 **a** $\frac{5}{3}$ m s⁻¹ **b** 1606.6 J
10 **a** $N = 3$ **b** $\frac{3}{8}$
11 0.3 m s⁻¹   3600 J
12 $v - u$ and $v + \frac{1}{2}u$

13 A moves with speed $1\,\mathrm{m\,s^{-1}}$ and B moves with speed $3\,\mathrm{m\,s^{-1}}$ in the same direction as before the impact.

14 a $2\,\mathrm{m\,s^{-1}}$ b $35\,\mathrm{J}$

15 $\dfrac{mu}{M+m}$

16 a $7.5\,\mathrm{m\,s^{-1}}$ b $375\,\mathrm{J}$ c $2\,\mathrm{kg}$

17 a $0.32\,\mathrm{s}$ $2.5\,\mathrm{m\,s^{-1}}$
    b $0.375\,\mathrm{J}$

Mixed exercise 4F

1 $26\,\mathrm{Ns}$, $23°$(nearest degree)

2 $-13\mathbf{j}\,\mathrm{m\,s^{-1}}$

3 a $61\,\mathrm{m\,s^{-1}}$ b $(68\mathbf{i}+23\mathbf{j})\,\mathrm{m\,s^{-1}}$

5 $\lambda = 4$

6 $u = \dfrac{-mV}{M}$

7 $v = 2.5\,\mathrm{m\,s^{-1}}$ $42.5\,\mathrm{J}$

8 $7\mathbf{i}+56\mathbf{j}$

9 c $\dfrac{3mu(1+e)}{4}\mathrm{Ns}$

10 $6\,\mathrm{m\,s^{-1}}$ and $1\,\mathrm{m\,s^{-1}}$ in the direction of the $100\,\mathrm{g}$ mass prior to the impact.
    Loss of K.E. $= 2.45\,\mathrm{J}$

11 $3\,\mathrm{s}$

12 As $\frac{5}{32}V < \frac{1}{4}V < \frac{7}{8}V$ there are no further collisions.

13 a $9072\,\mathrm{N}$ b *Either* heat *or* sound.

14 a $\frac{u}{7}(4-3e)$ and $\frac{4}{7}u(1+e)$

16 a Both balls change directions, the first moves up with speed $0.7\,\mathrm{m\,s^{-1}}$ and the second moves down with speed $3.5\,\mathrm{m\,s^{-1}}$.
    b 91% (2 s.f.)

17 a $0.5\,\mathrm{m}$ b $\dfrac{2}{\sqrt{g}}\mathrm{s}$ or $0.64\,\mathrm{s}$ (2 s.f.)
    c $\frac{1}{4}\sqrt{g}\,\mathrm{m\,s^{-1}}$ or $0.78\,\mathrm{m\,s^{-1}}$ (2 s.f.)

19 $\frac{25}{32}$

20 $V = \sqrt{\dfrac{2ME}{m(M+m)}}\,\mathrm{m\,s^{-1}}$

Exercise 5A

1 $1.09\,\mathrm{Nm}$ anticlockwise 2 $2.13\,\mathrm{Nm}$ anticlockwise
3 $1.31\,\mathrm{Nm}$ clockwise 4 $1.31\,\mathrm{Nm}$ clockwise
5 $1.31\,\mathrm{Nm}$ clockwise 6 $1.31\,\mathrm{Nm}$ clockwise
7 $1.78\,\mathrm{Nm}$ anticlockwise 8 $5.42\,\mathrm{Nm}$ anticlockwise
9 $3.60\,\mathrm{Nm}$ anticlockwise 10 $1.93\,\mathrm{Nm}$ anticlockwise

Exercise 5B

1 a $78.4\,\mathrm{N}$, $78.4\,\mathrm{N}$, $22°$ above AB
    b $45.7\,\mathrm{N}$, $45.7\,\mathrm{N}$, $40°$ above AB
    c $91.5\,\mathrm{N}$, $70.1\,\mathrm{N}$, parallel to AB
    d $32.4\,\mathrm{N}$, $32.4\,\mathrm{N}$, $115°$ above AB

2 a $18.4\,\mathrm{N}$ b $24.0\,\mathrm{N}$ c $27.6\,\mathrm{N}$ d $29.0\,\mathrm{N}$

3 $\dfrac{\sqrt{7}}{2}\,mg$, $49.1°$ to the horizontal

4 $88.3\,\mathrm{N}$, 5

5 $2.31\,\mathrm{m}$

6 a $16.3\,\mathrm{N}$ b $36.5\,\mathrm{N}$, $26.6°$ to the vertical

7 $22.9\,\mathrm{N}$, $22.9\,\mathrm{N}$, $40°$

8 $0.51\,\mathrm{m}$

9 a $13.9\,\mathrm{N}$ b $31.0\,\mathrm{N}$, $18.4°$ to the vertical

10 a $50.2\,\mathrm{N}$, $61.5\,\mathrm{N}$ b $1.95\,\mathrm{m}$ from A

Exercise 5C

1 a $2.69\,\mathrm{m}$ from A b $25.6\,\mathrm{N}$, $34.5\,\mathrm{N}$

2 a $38.1\,\mathrm{N}$ b $38.1\,\mathrm{N}$ at $40°$ to the rod

3 $28.3\,\mathrm{N}$, $139\,\mathrm{N}$ at $10.9°$ to the upward vertical

4 a $8.7\,\mathrm{N}$
    b $13.2\,\mathrm{N}$ at $19.1°$ to the upward vertical

5 $90.5\,\mathrm{N}$, $66.0\,\mathrm{N}$ at $75.96°$ to the upward vertical

6 a $29.3°$ to the vertical
    b $37.4\,\mathrm{N}$, $31.9\,\mathrm{N}$

Exercise 5D

1 $20\sqrt{3}\,\mathrm{N}$, $50\,\mathrm{N}$, $10\sqrt{3}\,\mathrm{N}$, 0.35

2 a $22.8\,\mathrm{N}$ b $98\,\mathrm{N}$, $22.8\,\mathrm{N}$ c 0.233

3 a $41.6°$ b $24.0°$

4 $5\frac{1}{3}\mathrm{m}$

5 a $\dfrac{8\sqrt{5}}{9}\mathrm{N}$ b 0.526

7 $104\,\mathrm{N}$, $64.5\,\mathrm{N}$, 0.620

9 a $\dfrac{W}{2\sqrt{3}}$ c $\dfrac{W}{4}$

10 a $\dfrac{15\sqrt{3}}{4}\mathrm{N}$, $\dfrac{25\sqrt{3}}{4}\mathrm{N}$ b $\dfrac{1}{\sqrt{3}}$

11 $50.2°$

12 $59.0°$

Mixed exercise 5E

1 $150\,\mathrm{N}$

2 $\dfrac{5Mg}{12}$

3 $1.5\,\mathrm{N}$, $6.4\,\mathrm{N}$

4 $21.8°$

6 a $\dfrac{13W}{10}$ b $\dfrac{6W}{5}$

7 $\frac{7}{24}$

8 $743\,\mathrm{N}$

9 $0.39l$

10 a $\dfrac{8W}{9}$
    c The ladder has negligible thickness/the ladder does not bend.

11 a $830\,\mathrm{N}$ b $620\,\mathrm{N}$

12 b $\dfrac{\sqrt{10}W}{2}$

13 c $\dfrac{W}{4g}$

14 $7470\,\mathrm{N}$ at $48°$ to the horizontal

Review Exercise 2

1 a $(10\mathbf{i}+20\mathbf{j})\,\mathrm{m\,s^{-1}}$ b $63.4°$ c $40\,\mathrm{J}$

2 a $7.5\,\mathrm{N}$ b $(39\mathbf{i}-42\mathbf{j})\,\mathrm{m\,s^{-1}}$

3 a $5.83\,\mathrm{Ns}$ b $31°$ (nearest degree)
    c $35\,\mathrm{J}$

4 a $\mathbf{v} = (2t+2)\mathbf{i}+(3t^2-4)\mathbf{j}$
    b $13\,\mathrm{m\,s^{-1}}$

5 a $(-25\mathbf{i}-5\mathbf{j})\,\mathrm{m\,s^{-1}}$
    b $32.9\,\mathrm{m}$ c $51\,\mathrm{m}$

6 a A has speed $\dfrac{u}{2}$ B has speed $\dfrac{3u}{2}$
    b $\dfrac{45}{2}mu^2$

7 a $\frac{1}{4}$ b $\dfrac{9mu^2}{4}$

8 $\frac{3}{5}$

10 a $v_B = \frac{2u}{5}(1 + e)$ $v_A = \frac{u}{5}(2 - 3e)$

   c $e = \frac{5}{6}$

11 a $\frac{u}{6}(5e - 1)$ **b** $\frac{1}{5} < e \le 1$

12 b $\frac{1}{4}(1 - 3e)u$

   c $e = \frac{1}{3}$

13 b $\frac{25}{4}mu^2$

14 b i $\frac{1}{3}u(2e - 1)$

   ii The direction of motion was reversed.

15 b $\frac{2}{3} < e \le 1$ **c** $\frac{7}{9}$

16 a $v_B = \frac{1}{4}u(1 + e)$ $v_A = \frac{1}{4}u(3e - 1)$

   b $\frac{1}{3} < e < \frac{7}{9}$

17 a $\frac{2}{3}$

   b As speed A > speed B there will be no further collisions.

18 b $\frac{25}{32}$

   c Q is now moving towards the wall once more. After Q hits the wall, it will return to collide with P once more.

19 c $\frac{1}{9} < f \le 1$

20 b $v_A = \frac{1}{5}u(7 - 3e)$

   d $\frac{5d}{16}$

21 a uniform rod

23 b $X = \frac{2}{3}W$

24 b $mg\sqrt{13}$

25 b $Y = \frac{5mg}{4}$

   c The tension would not be constant throughout the length of the string.

26 b $0.4mg \le P \le 13.6mg$

27 a $mg\frac{\sqrt{3}}{3}$

28 a

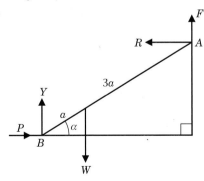

   b $P = \dfrac{W}{4\mu + 8}$

29 b 68.2°   **c** 0.646

30 $\frac{1}{4}$

31 a $S = \frac{7W}{8}$

   c The ladder will be straight.

32 $0 \le x \le \frac{11}{4}$

33 c $\frac{1}{4}M$

   d

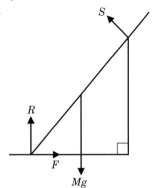

34 a 28.3 N   **b** 74.8 N

35 a $T = 1020$ N   **b** 778 N

   c So for equilibrium, the moment of R must be zero. Hence R must pass through M and so is horizontal.

36 $\theta = 63°$ (nearest degree)

37 Tension = $\dfrac{5W}{2\sin\theta}$,

   Vertical = $\dfrac{W}{2}$,

   Horizontal = $\dfrac{5W}{2\tan\theta}$

39 a 94 N

   b 0.47 (2 d.p.)

   c 133 N (nearest N)

40 b $mg\dfrac{\sqrt{5}}{2}$

   c In the moments equation.

Practice Exam Paper

1 a 7.8 N (2 s.f.)

   b 14 N (2 s.f.)
   71° (2 s.f.)

2 8.1 m s^{-1} (2 s.f.)

3 a i $\frac{7a}{5} = \bar{x}$ **ii** $\frac{4a}{5} = \bar{y}$

   b 30°

4 a 14 m s^{-1} **b** $\frac{30}{7}$ s

   c 21 m s^{-1} **d** 53° (nearest degree)

6 a $t = 0$ or 3   **b** 44.75 m

7 a 11 500 N

   b 6.2 m s^{-2}

   c 3700 N (2 s.f.)

Index

acceleration
  constant 2–11, 112
  gravity 64
  Newton's Second Law 15, 22, 65, 78
  varying with time 11–19
  vectors 13, 19–23
  work 63–4
accuracy of data 3
angle of elevation 2
angle of projection 2
approach speed 105–7, 111–12, 115, 117, 121
arcs 50–1
area of triangle formula 44
axes of symmetry 39–41, 45, 50

balls, elasticity 105

calculus 11–25
cantilevers 131
centres of mass 31–62
centroid of triangle 40
circles 39–52
coefficient of friction 53
coefficient of restitution (e) 104–15, 117–18, 121
collisions 100–30
composite laminas 44–50
concurrent forces 140
conservation of linear momentum 101–10, 115, 121–2, 124
conservation of mechanical energy 72–6
constant acceleration formulae 2–11, 112
constant of integration 20–1, 23

differentiation 11–17, 19–23
discs 39, 47–8
displacement 11–17

distance
  kinematics 2, 5, 11
  work 64–7, 74, 77
dots, differentiation 19

e (coefficient of restitution) 104–12, 117–18, 121
elasticity 105
energy 63–99, 121–6
engines, power 77–81
equilibrium
  laminas 53–8
  limiting 141–6
  rigid bodies 134–46
exam practice paper 161–3

flight time of projectiles 2
forces
  friction 53, 73, 141–3
  impulse–momentum principle 101
  moments 32, 35, 132–6, 139, 142–3
  Newton's Second Law 15, 22, 65, 78
  non-parallel 139–41
  pulling 77
  rigid body statics 132–46
  tractive 77
  work 64–7, 74, 77
formulae
  area of triangle 44
  centres of mass 32–9, 44–50
  constant acceleration 2–11, 112
  distance 2, 5, 11
  energy 68, 122
  kinetic energy 122
  momentum 101, 107–8
  Newton's Law of Restitution 104–15, 117–18, 121
  Newton's Second Law 15, 22, 65, 78

power 77–9
sum of infinite GP 119
trigonometry 7
vector magnitude 22
work 64–8, 74, 77
frameworks, centres of mass 50–2
friction 53, 73, 141–3

geometric progressions (GPs) 119
gravity 1
  collisions 112, 119
  conservation of mechanical energy 72–3
  constant acceleration formulae 112
  work done 64–5
  see also forces

horizontal component, velocity 2–8

i and j vector components 8, 35, 37–8
impacts 100–30
impulse–momentum principle 100–4, 107–8, 123
inelastic particles 105
integration 11–17, 19–25

j and i vector components 8, 35, 37–8
joules 64, 77, 124

K.E. (kinetic energy) 68–72, 121–6
kinematics 1–30
kinetic energy (K.E.) 68–72, 121–6

laminas 39–58, 132
limiting equilibrium 141–6

magnitude
    moments 132
    vectors 22
mass
    centres of 31–62
    conservation of linear
        momentum 101
    Newton's Second Law 15, 22,
        65, 78
mechanical energy 72–6
medians of triangles 40
modelling assumptions,
    kinematics 1
moments
    centres of mass 32, 35
    rigid body statics 132–6, 139,
        142–3
    see also forces
momentum 100–10, 115, 121–4
moving vehicles 77–81

Newton-metres 132
Newton-seconds 101
Newtons 64, 77
Newton's Experimental Law 104
Newton's Law of Restitution
    104–15, 117–18, 121
Newton's Second Law 15, 22,
    65, 78

P.E. (potential energy) 68–72
perfectly elastic particles 105,
    127
planes 1–30
    centres of mass 34–9
    collisions 111–21
    constant acceleration 2–11
    kinematics 1–30
    vertical movement 2–11
position vectors 19–23, 35, 45
potential energy (P.E.) 68–72
power 63–99
practice exam paper 161–3
projectiles 1–11

pulling (tractive) forces 77
Pythagoras' Theorem 8, 21, 102,
    135

quadratic equations 6
quadratic formula 7

range, projectiles 2
rebound speed 111–12, 117–18
rectangular laminas 39–52
resolving forces 142–3
restitution
    coefficient 104–15, 117–18,
        121
    Newton's Law 104–15,
        117–18, 121
resultant forces 134–5, 140
review exercises 86–99, 151–60
rigid body statics 131–50
rods 40, 50–2
rotation, moments 132

sectors of circles 41
separation speed 105–7, 111,
    115, 121
simultaneous equations 108,
    115, 117, 121
Skywalk platform, Grand
    Canyon 131
speed
    approach 105–7, 111–12, 115,
        117, 121
    power 77
    projection 2
    rebound 111–12, 117–18
    separation 105–7, 111, 115,
        121
statics of rigid bodies 131–50
straight lines 1–30, 32–4
suspended laminas 53–4
symmetry
    axes 39–41, 45, 50
    collisions 118

time of flight, projectiles 2
tractive (pulling) force 77
triangle of forces 139–41
triangular laminas 39–50, 44–5
trigonometry
    centres of mass 55
    impulse–momentum
        principle 102
    kinematics 5, 7–8
    potential energy 70
    rigid body statics 135

uniform laminas 39–58

vectors
    acceleration 13, 19–23
    constant of integration 20–1,
        23
    $\mathbf{i}$, $\mathbf{j}$ components 8, 35, 37–8
    magnitude formula 22
    momentum 101–4
    position vectors 19–23, 35,
        45
    triangle of forces 140
    velocity 19–23, 123
vehicle movement 77–81
velocity
    collisions 101, 107
    conservation of linear
        momentum 101
    kinematics 2–8, 11–17, 19–23
    vectors 19–23, 123
vertical component, velocity
    2–8
vertical planes 2–11

watts 77
work 63–85
    change in kinetic energy 68
    done against gravity 64–5
    force at angle to motion 66
    force × distance 64–7, 74, 77
    power 77–81
    work–energy principle 72–6

A Level Academy
Guildford College
Stoke Park
Guildford
Surrey
GU1 1EZ